Environmental Quality, Innovative Technologies, and Sustainable Economic Development

A NAFTA PERSPECTIVE

Proceedings of a Workshop

Sponsored by the

National Science Foundation, United States
Sandia National Laboratories, United States
Consejo Nacional de Ciencia y Tecnología, México
National Science and Engineering Research Council, Canada

Organized by the

Center for Sustainable Technology, Georgia Institute of Technology
Instituto de Ingeniería, Universidad Nacional Autonoma de México
Department of Civil Engineering, University of Western Ontario

Mexico City, Mexico
February 8-10, 1996

Approved for publication by the
Environmental Engineering Division of the American Society of Civil Engineers

Edited by
Emir José Macari
F. Michael Saunders

Published by

ASCE American Society of Civil Engineers
345 East 47th Street
New York, New York 10017-2398

Abstract:

This workshop proceedings assesses the current level of environmental technologies available in key media and industrial areas and their role in enhancing environmental quality in international marketplaces. It addresses topics such as: 1) Sustainable development and industrial ecology; 2) water and waste at the U.S./Mexico border; 3) earth system science for sustainable development of earth resources; 4) air pollution engineering; 5) integration of biotechnology in remediation and pollution prevention activities; and 6) geo-environmental concerns in North America. Discussions such as these encourage engineers to develop the broadened view of technology, environmental awareness, and a deep commitment to serve society.

Library of Congress Cataloging-in-Publication Data

Environmental quality, innovative technologies, and sustainable economic development : a NAFTA perspective : proceedings of a workshop : Mexico City, Mexico, February 8-10, 1996 / edited by Emir José Macari, F. Michael Saunders ; sponsored by the National Science Foundation, United States...[et al.] ; organized by Center for Sustainable Technology, Georgia Institute of Technology, Instituto de Ingeniería, Universidad Nacional Autónoma de México, Department of Civil Engineering, University of Western Ontario.
p. cm.
Includes index.
ISBN 0-7844-0224-8
1. Environmental sciences--Research--North America--Congresses. 2. Pollution--Environmental aspects--North America--Congresses. I. Macari, Emir José. II. Saunders, F. Michael. III. National Science Foundation (U.S.)
GE90.N7E59 1997 97-157
363.7'0072073--dc20 CIP

Library of Congress Catalog Card No: 97-157
ISBN 0-7844-0224-8
Manufactured in the United States of America.

ACKNOWLEDGEMENTS

This workshop proceedings is the result of hard work from many individuals who are committed to uniting the environmental scientific and technological communities of North America. Special recognition is first offered to Dr. Eleonora Sabadel of the U.S. National Science Foundation who encouraged us all to hold this meeting at a critical time in history for this region of the world. Her vision and insight greatly contributed to the success of this meeting.

The organizing committee was composed of members from each of the three NAFTA countries. Dr. Adalberto Noyola, Director of the Environmental Bioremediation group of the Institute of Engineering, UNAM and Dra. Mariza Mazari of the Institute of Ecology where the perfect hosts to this event. Their hard work and dedication was evident as every logistical aspect of the meeting went untarnished. The venue location, the University's (UNAM) Botanical Garden, proved to be a very conducive setting for this environmental meeting. On the part of Canada, Dr. Kerry Rowe, Professor of Civil Engineering from the University of Western Ontario believed in the objectives of this meeting from the very beginning and never gave up to make sure that the Canadian environmental community would be well represented. Drs. James Mulholland, and J. David Frost were key members of the U.S. organizing committee. From the preparation of the proposal seeking funds for this meeting to ensuring that the most qualified members of the environmental community of the U.S. participated, the both were closely involved at every step. We would also like to thank Dr. Dennis Engi of the Sandia National Laboratoies (DOE) for helping fund the preliminary work that led to this workshop. His participation in this initiatives were most helpful.

This workshop addressed issues of environmental concern that affect North America. In addition, the workshop united a group of environmental experts from the three NAFTA countries in order to begin working in cooperation to solve these problems that, for the most part, transcend national boundaries. These proceedings constitutes the summaries of discussions during the breakout sessions as well as a compendium the position papers that were presented by the participants prior to the meeting. The underlying objective of the workshop was to promote cooperative programs between industry, government, and academia in North America. The organizing committee would like to express their sincere appreciation to all of the workshop participants. Their hard work over the three long days and nights of meetings has resulted in this proceedings which should prove to be an excellent reference for environmental engineers and scientist, environmental policy makers, and environmental companies interested in a detail insight of the problems and potential solutions to many of the environmental problems that affect our region. ASCE's Environmental Engineering Division Technical Committee is gratefully acknowledged for reviewing and endorsing this document.

Emir José Macari and F. Michael Saunders, Co-Editors

FOREWORD

Contamination of soil, air and groundwater are problems common to developed as well as developing countries; the impacts on health and the deterioration of natural resources transcends geographic and economic boundaries. Environmental conservation and protection has often taken a back seat to economic development. This has led to increasing concerns on how to reduce waste and mitigate problems caused by past production activities in order to ensure that the earth remains a safe haven for future generations. The 1992 Earth Summit in Rio de Janeiro was a first step in addressing environmental protection as a global problem. Since then, an increasing amount of attention has been devoted to the concept of sustainability through research and development initiatives.

Innovative technologies have the potential to address current issues related to environmental quality in air, land, and water media and also are critical to establishing the foundation for sustainable economic development in all nations of the world. Developed and developing countries need to assess the role of current and emerging innovative technologies in addressing current and projected environmental problems; developing countries need to gain insight from the experiences of those countries that have already traveled the path to economic development and avoid the inappropriate or non-sustainable approaches of the past. To assist in this effort, the workshop focused on the gathering of an international forum of individuals from Mexico, Canada, and the U.S. to identify critical research needs and set priorities and approaches for the pursuit of these initiatives from a perspective of the recently signed North American Free Trade Agreement. The purpose of the workshop was, thus, to **assess the current level of environmental technologies available in key media and industrial areas** and to **assess their role in enhancing environmental quality in international marketplaces**. The topic areas addressed included **air, water, and land media**, as well as **environmentally-conscious manufacturing, environmental health risk assessment, environmental regulations, and environmental technology transfer**.

In December, 1994, President Clinton hosted the first Hemispherical Summit in Miami, Florida. The main topics of the summit were Democracy and Economic Sustainable Development. The recognition of the interdependence between the environment, technology and economic development, is leading to significant changes in how systems will be engineered in the future. Achieving a high quality of life without environmental degradation requires a transformation in technology and its management.

Reports have indicated that NAFTA has opened up new opportunities for American businesses in environmental cleanup in Mexico. Carol Browner, EPA Administrator, has stated that with NAFTA, "we are beginning to see a new era of environmental protection." She went on to say that the United States has begun to see increased sales of environmental protection services to Mexico. The World Bank has announced the approval of $368 million for water, sewage, and air pollution projects in Mexico. In addition, an estimated $800 million will be available over the next three to four years. These announcements have stressed the need for environmental experts from NAFTA countries to sit together to discuss strategies and prioritize projects that will support the trade efforts ensuring that environmental conservation and remediation become a high priority item of any negotiation.

Gabriel Quadri de la Torre, president of Mexico's National Institute of Ecology stated recently (October, 1996) that Mexico needs nearly $10 billion per year to solve their environmental problems. He stated that this investment is needed to develop the infrastructure necessary to solve some of the most pressing environmental problems in Mexico such as waste water management, toxic waste management and control of atmospheric contamination. He also said that the new Mexican Council of Environmental Investment will be charged with generating new projects and as well as informing prospective investors of the opportunities in the area of environmental control and remediation.

Worldwide, in 1994, the environmental market represented some $408 billion dollars of which 88% was attributed to the United States, western Europe and Canada, where priority is given to investment in waste water (39.4%), equipment and management of solid waste 28.2%, and environmental consulting and equipment for pollution control 12.7%.

Mr. Quadri de la Torre explained that Mexico will need approximately $6.85 billion dollars from 1997-2000 for the management of waste water and another $2.5 billion dollars for treatment of industrial waste water. It is thought that if proper investments are made now by the year 2010 Mexico will have adequate infrastructure to solve most of the various environmental problems. He said that in urban zones, an estimated 30 million metric tons of waste are generated per year which is equivalent to 329 kg. per person per year.

The newly enacted environmental regulatory reform in Mexico has set as main objective to instill in its law the trends of the new environmental policy which is founded on the principles of sustainable development. Mr. Gabriel Quadri de la Torre presented a very inspiring keynote lecture in which he stated the philosophy his agency has followed in this regulatory reform effort. In addition, he answered questions from the audience as well as listened to suggestions from the forum of environmental experts on recent global trends on environmental regulatory policies.

The second keynote speaker, Dr. Brad Allenby, Director of Energy and Environmental Systems from the Lawrence Livermore National Laboratory presented his perspectives and philosophy regarding environmental protection in the context of Industrial Ecology. Many of his remarks are presented in his position paper which is the result of a report to the Institute of Electrical and Electronic Engineers on Sustainable Development and Industrial Ecology.

The third keynote speaker, Dr. William S. Fyfe, Professor of Environmental Sciences of the University of Western Ontario presented his views on the protection of natural resources from a much more global perspective. His ideas are well detailed in his position paper titled "Earth System Science for Sustainable Development of Earth Resources."

These three keynote presentations set the stage for the discussions that would take place over the next three days. The focus of these discussions were be on environmental issues in North America but not forgetting that this region is only one part of a larger continent.

The signing of the NAFTA environmental side agreement resulted in the creation of the North American Commission of Environmental Cooperation (CEC) which has set as a mission to facilitate the cooperation and public participation to foster conservation, protection and enhancement of the North American environment for the benefit of present and future generations, in the context of increasing economic, trade and social links among Canada, Mexico and the United States.

The original intention of NAFTA was to begin to form an alliance between Mexico, Canada and the United States with the future hope of expanding a free trade zone not only within North America but between the North Slope of Alaska to the tip of Tierra del Fuego. It is widely known that the passage of the North American Free Trade Agreement has resulted in expanded market opportunities for U.S. and Canadian environmental service firms in Mexico. However, recent trends in environmental protection and restoration in Central and South America have widened the market opportunities for all three NAFTA countries. Since August, 1996, the Central American governments have strengthen their efforts in environmental protection projects through a regional Sustainable Development Alliance (Alianza para el Desarrollo Sostenible, ALIDES). Environmental meetings designed to help set policies and the course to follow increased during the summer of 1996. These efforts are fully endorsed by the presidents of the entire region as evidenced in the discussions that took place during the last two presidential summits in December, 1995 and May, 1996. NAFTA countries have joined together to become extraregional sponsors of ALIDES. Firms that have set up partnerships in Mexico have had ample success in exporting the technologies to the southern countries. Cultural and regional differences have been minimized by these partnerships and everyone has emerged as a winner.

The second Summit of the Americas will take place in Bolivia in December, 1996. It is expected that one of the key items in the agenda will be the promotion of environmental protection and restoration of our Hemisphere. Participants will likely endorse the use of renewable energy sources in Latin America and seek a regional commitment to sustainable development policies.

NAFTA and its environmental side agreement has been the subject of analysis and controversy since its inception. However, one thing is clear, environmental awareness in Latin America has increased as a result of such agreements. A recent book titled "The Environment and NAFTA: Understanding and Implementing the New Continental Law," by Johnson and Beaulieu places many of these issues into perspective. The book points out the good working relationship that the Ministers of the Environment for the three NAFTA countries have had. The Council of Ministers is composed of Canadian Environment Minister Sergio Marchi, the Mexican Secretary of Environment, Natural Resources and Fisheries, SEMARNAP, Julia Carabias, and U.S. Environmental Protection Agency, EPA, Administrator Carol Browner. Action plans aimed at reducing the dangerous effects of PCBs, mercury, chlordane and DDT in North America are being proposed by this council and an agreement is expected to be ratified before the end of 1996.

The agenda set forth by President Clinton in his science and technology policy document, "Science in the National Interest" in regard to cooperation within our Hemisphere have begun to pay dividends. "As a result of deliberate and successful long-term investment strategies, a number of countries now possess world-class research capabilities. If U.S. researchers are to sustain leadership and strengthen participation in collaborative scientific endeavors, we must increase our level of interaction with colleagues in other countries. In many important areas of contemporary research, ranging from studies of seismic activity to biodiversity to global change, our scientists can be optimally effective only through international partnerships." "We should also look for opportunities to engage developing nations more fully in the international science endeavor. As a logical consequence of the North American Free Trade Agreement and (other) long-term policies, we should continue to pay particular attention to engaging the scientific communities of the Americas."

Tremendous opportunities for protecting our environment, in conjunction with strong economic development, lie ahead as we move into the 21st century. Sustainable Development is a key to ensuring that future generations will have the same opportunities and natural resources at their disposal as we do today. Environmental health risks issues must be addressed for the good of our children and their children and the only way to ensure that our Hemisphere is safe for those future generations is to act now. We must begin to produce a new breed of engineers and scientist with a broadened view of technology, environmental awareness, and a deep commitment to serve our society. This workshop promotes these ideas and policies.

Emir José Macari, Chair

TABLE OF CONTENTS

Workshop Objectives

The objectives of this workshop were twofold: first, to unite experts from the United States, Canada, and Mexico, thereby fostering international cooperative research programs in environmental technology in support of the North American Free Trade Agreement; and second, to identify and prioritize research areas that can contribute to the first objective. For the purposes of logistics and organization, the umbrella topic of Environmental Technologies has been divided into the following areas:

Water

In the water area, issues to be investigated include planning and management of water resource systems focusing on regional basins, as well as international watershed, management and quality and use of saline and brackish waters as sustainable water resources. Issues of water quality for potable and industrial use include microbial contamination (e.g., E. coli, Giardia, Cryptosporidium and viral particles), inorganic contamination (e.g., arsenic and selenium, lead, mercury, nitrite/nitrate, ammonium, phosphates) and organic contamination (e.g., domestic wastes, pesticides, organic solvents). The current and emerging technologies in these areas will be examined in an integrated setting to assess appropriate points and levels of applications.

Soil and Subsurface

The cost of environmental restoration of contaminated soil varies from $50 per cubic meter of soil to about $1500 per cubic meter in the United States. It has become evident that such costs are very detrimental to economic growth, given the extent of subsurface contamination. The discipline of geo-environmental engineering has evolved over the past two decades to address issues related to waste containment systems; site investigation, characterization, evaluation, and monitoring; risk assessment and minimization; remediation systems for contaminated sites; performance assessment, quality assurance, and quality control; site closure of site improvement; reuse and recycling of waste materials; pollution prevention; and sustainable geo-environmental technologies. Notwithstanding important development in techniques and technologies for dealing with subsurface contamination issues, it is evident that, as in other fields, the concept of sustainability has not always been a key consideration in developments. Given the new perception of the interdependence between the environment, technology and economic development, it is thus important that the field of geo-environmental engineering accelerate efforts to study ways in which existing systems can be modified and new systems be engineered in the near future.

Air

Air quality remains an important are for cooperative study due to the international nature of many global and regional issues and an improved understanding of the environmental health risks. Global warming and stratospheric ozone depletion are problems of vast scale. Acid deposition and photochemical smog are examples of regional and local air pollution problems. New technologies are being developed for monitoring, controlling, and preventing emissions of greenhouse gases, stratospheric ozone destroyers, oxides of sulfur and nitrogen, particulate matter, unburned hydrocarbons, and air toxins such as poly-aromatic compounds and metal aerosols. These developments will need the cooperation of the international community in order to make their impacts applicable to local, regional, and international settings.

Environmentally Conscious Products, Processes and Manufacturing
A progressive assessment of current and pending environmental regulation s on commercial and industrial operations has resulted in a paradigm shift away from a simple and rigid compliance-only approach to a design for the environment (DFE) approach. DFE engenders a systematic consideration and evaluation during new product and process development of design issues associated with environmental safety and health over the full life cycle of materials and products. DFE results in a focus on both enterprise integration for agile manufacturing and on sustainable development making industrial growth and environmental quality mutually reinforcing and compatible.

Environmental Health Risk
The effective development, examination and use of current and emerging technologies requires that environmental risks be anticipated, well beyond those engendered in current compliance and environmental regulations. This requires an ability to examine material for manufacture and products throughout the full life cycle and anticipate their ultimate fate in the environment and the associated human and environmental exposure and health risks.

A recent idea that the U.S. National Institutes of Health are promoting is the collaboration of engineers with public health scientists to provide better estimates of health risks due to exposure to a wide range of pollutants. For example, we need better markers of health risk, such as human dosimeters for the types of pollutants at levels consistent with human experience (i.e., very low exposures). There needs to be a close interaction between these developing technologies and those estimating health risk, etc.

AGENDA

WORKSHOP ON

"Environmental Quality, Innovative Technologies, and Sustainable Economic Development - A NAFTA Perspective"

Mexico City, Mexico
February 7-10, 1996

Day 1. February 7.

19:00	Welcoming Cocktail.
	Honorary Guests:
	The Honorable Julia Carabias, Secretary of the Environment for Mexico (SEMARNAP)
	Canada Ambassador to Mexico, the Honorable Marc Perron
	U.S. Ambassador to Mexico, the Honorable James Jones
	Dr. José Sarukhan, Chancellor of the National University of Mexico

Day 2. February 8.

9:00 Welcome by Dr. José Luis Fernández Zayas, Institute of Engineering, UNAM

9:00 - 10:30	**Plenary Session - Keynote Presentations**
	Ing. Gabriel Quadri de la Torre, *Mexico*
	President of the National Institute for the Ecology
	Dr. Brad Allenby, *United States*
	Director for Energy and Env. Systems, Lawrence Livermore National
	Dr. William S. Fyfe, *Canada*
	Director, Department of Earth Sciences, University of Western Ontario
10:30 - 11:00	Coffee Break
11:00 - 11:30	**Workshop Objectives and Organization.** Macari/Noyola
11:30 - 13:30	Break-out sessions 1, 2, and 3 **
13:30 - 15:00	Break / Lunch
15:00 - 16:00	Discussion and Conclusions of Break-out sessions 1, 2, and 3
16:00 - 17:00	Plenary session. Summary presentations of Break-out sessions 1, 2, and 3.
17:00 - 17:15	Organization of Day 3.

Day 3. February 9.

9:00 - 11:00	Break-out sessions 4, 5, and 6 **
11:00 - 11:30	Coffee Break
11:30 - 12:30	Discussion and Conclusions of Break-out sessions 4, 5, and 6
12:30 - 13:30	Plenary session. Summary Presentations of Break-out sessions 4, 5, and 6.
13:30 - 15:00	Break / Lunch
15:00 - 17:00	Break-out sessions 7, 8, and 9 **
17:00 - 18:00	Discussion and Conclusions of Break-out sessions 7, 8, and 9
18:00 - 19:00	Plenary Session. Summary Presentations of Break-out sessions 7, 8, and 9

Day 4. February 10.

9:00 - 10:00	Plenary Session. General Summary of Findings from Break-out sessions
10:00 - 12:00	General Discussion of Findings.
	Responsibilities, Follow-up, and Accountability,
12:00	Workshop ends

**** Break-out Sessions:**

1. **Groundwater**
2. **Air pollution (Local and Regional)**
3. **Environmental Health Risk Assessment**
4. **Waste water treatment**
5. **Solid Wastes (Municipal and Industrial)**
6. **Legal and Regulatory Environmental Issues**
7. **Soil Remediation**
8. **Environmentally Conscious Manufacturing**
9. **Technology Transfer, Training and Intellectual Property Rights**

"Environmental Quality, Innovative Technologies, and Sustainable Economic Development - A NAFTA Perspective"

Participant Information

MEXICO

ANGELES, ANTONIO — SECRETARIA DEL MEDIO AMBIENTE
PLAZA DE LA CONSTITUCION #1, TERCER PISO
PHONE: 5-10-47-37 EXT. 1404
FAX: 5-10-47-37 EXT. 1404

AUVINET, GABRIEL — INSTITUTO DE INGENIERIA, UNAM.
CIUDAD UNIVERSITARIA
PHONE: 622-3500
FAX: 616-0784
gag@jazz.iingen.unam.mx

CARMONA, M. DEL CARMEN — INSTITUTO DE INVESTIGACIONES JURIDICAS, UNAM
CIUDAD UNIVERSITARIA
PHONE: 622-7463, 622-7477

BUITRON, M. GERMAN — INSTITUTO DE INGENIERIA, UNAM
CIUDAD UNIVERSITARIA, A.P. 70-472
C.P. 04510, MEX., D.F.
PHONE: 6-22-33-24
FAX: 6-16-21-64
gbm@pumas.iingen.unam.mx

DOWNS, TIM — INSTITUTO NACIONAL DE SALUD PUBLICA/U.C.L.A.
AV. UNIVERSIDAD #655, CUERNAVAVA, MORELOS
C.P. 62508
PHONE: 73-110111 / 2656
FAX: 73-11-11-48

ESPINOSA, GABRIELA — INSTITUTO MEXICANO DEL PETROLEO
EJE CENTRAL LAZARO CARDENAS #152, SN BARTOLO
PHONE: 368-3788
FAX: 567-2928

ESPINOSA, JAVIER A. — UNIVERSIDAD DEL VALLE DE MEXICO
HDA. MIMIAHUAPAN #54, COL. HDA. EL ROSARIO,
C.P. 02720.
PHONE: 3-82-70-74

ESTEVA, J. ANTONIO — CENTRO PARA LA INNOVACION TECNOLOGICA
APARTADO POSTAL 20-103, CIUDAD QUNIVERSITARIA, MEXICO
PHONE: 550-9192, 622-5204
FAX: 622-5223
maraboto@servidor.unam.mx

GUZMAN, FRANCISCO	INSTITUTO MEXICANO DEL PETROLEO EJE CENTRAL #152, MEXICO, DF. C.P. 07730 PHONE: 5-67-92-46 FAX: 5-87-00-09 pancho@tzoalli.sgia.imp.mx
JAIME, ALBERTO	INSTITUTO DE INGENIERIA, UNAM COMISION FEDERAL DE ELECTRICIDAD MELCHOR OCAMPO #469, NOVENO PISO, C.P. 11590, MEXICO, D.F. PHONE: 254-76-72 FAX: 254-7035
KURI-HARCUCH, W.	INVESTIGACIONES Y ESTUDIOS AVANZADOS INSTITUTO POLITECNICO NACIONAL AV. IPN #2508, COL. SAN PEDRO ZACATENCO PHONE: 747-7000 EXT. 5521 FAX: 747-7081 walid@cell.cinvestav.mx
LESSER J.,MANUEL	LESSER Y ASOCIADOS, S.A. DE C.V. RIO GUADALUPE #3, COL. PATHE, C.P. 76020 QUERETARO, MEXICO PHONE: 91 (42) 233361 FAX: 91 (42) 231515
MARTIN, PAUL	BANCO MUNDIAL (WORLD BANK) PLAZA INN #71, INSURGENTES SUR 1971, MEXICO, D.F. PHONE: (525) 661 1791 FAX: (525) 661-0917 pmartin@worldbank.org
MAZARI, MARISA	CENTRO DE ECOLOGIA, UNAM. CIUDAD UNIVERSITARIA, APARTADO POSTAL 70-275 C.P. 04510, MEXICO, D.F. PHONE: 6-22-89-98 FAX: 6-22-89-95 Y 6-16-19-76 mazari@servidor.unam.mx
NOYOLA R., ADALBERTO	INSTITUTO DE INGENIERIA, UNAM. APARTADO POSTAL 70-472, CIUDAD UNIVERSITARIA, COYOACAN, C.P. 04510, MEXICO, D.F. PHONE: (52-5) 6-223326, 6-223327 FAX: (52-5) 6-162164 noyola@pumas.iingen.unam.mx
OLGUIN, EUGENIA J.	INSTITUTO DE ECOLOGIA APARTADO POSTAL 63 XALAPA, VERACRUZ, C.P. 91000, MEXICO PHONE: (28) 18-60-00 / 18-62-09 EXT. 4400 FAX: (28) 18-78-09 eugenia@sun.ieco.conacyt.mx

OLIVARES, J. MANUEL

PEMEX REFINACION
MARINA NACIONAL #329, MEXICO, D.F.
PHONE: 250-66-64
FAX: 627-76-71

ORTIZ V., ALEJANDRO

SECRETARIA DEL MEDIO AMBIENTE
PLAZA DE LA CONSTITUCION #1, TERCER PISO
PHONE: 5-10-47-37 EXT. 1404
FAX: 5-10-47-37 EXT. 1404

OSTROSKY, PATRICIA

INSTITUTO DE INVESTIGACION BIOMEDICA, UNAM.
CIRCUITO INTERIOR. ENTRE MEDICINA Y QUIMICA
CIUDAD UNIVERSITARIA
PHONE: 622-3846
FAX: 622-3846, 550-0048
ostrosky@servidor.unam.mx

REVAH, SERGIO

UNIV. AUTONOMA METROPOLITANA-IZTAPALAPA.
DEPTO. DE INGENIERIA DE PROCESO, AV.
PURISIMA S/N, COL. VICENTIRIA, C.P. 09340,
MEXICO, D.F.
PHONE: (5) 724-46-48 al 51
FAX: (5) 724-49-00
srevah@xanum.uam.mx

ROSALES A., EFRAIN

AMCRESPAC/MULTIQUIM
CAPIROTE #34, COL. SAN LORENZO HUIPULCO,
D.F., MEXICO
PHONE: 573-72-82
FAX: 655-36-15

RUIZ, MA. ESTHER

INSTITUTO MEXICANO DEL PETROLEO
AV. 100 METROS #152
PHONE: 567-8599
FAX: 587-0009
marisa@tzoalli.sgia.imp.mx

SALAZAR, JAQUELIN

INSTITUTO MEXICANO DEL PETROLEO
EJE CENTRAL LAZARO CARDENAS #162
PHONE: 368-5911 EXT. 21424, 20635

SANTOS, CARLOS

INSTITUTO NACIONAL DE SALUD PUBLICA
RIO SINALOA #318, CUERNAVACA, MORELOS, C.P.
62290
PHONE: 91-73-153217/113783
FAX: 91-73-111156

SAVAL, SUSANA

INSTITUTO DE INGENIERIA, UNAM
APARTADO POSTAL 70-472, CIUDAD
UNIVERSITARIA, C.P. 04510, MEXICO, D.F.
PHONE: 622-33-24 AL 29
FAX: 616-2164

VALDES, CARLOS
INSTITUTO MEXICANO DEL PETROLEO
EJE CENTRAL LAZARO CARDENAS #152, COL. SAN BARTOLO ATEPEHUACAN, MEXICO, D.F.
PHONE: 368-9212
FAX: 368-9212

CANADA

ROWE, R. KERRY
UNIVERSITY OF WESTERN ONTARIO
DEPARTMENT OF CIVIL ENGINEERING,
LONDON, ON CANADA N6A 589
PHONE: (519) 661-2126
FAX: (519) 661-3942
kerry@engga.uwo.ca

SHORTREED, JOHN H.
UNIVERSITY OF WATERLOO
INSTITUTE FOR RISK RESEARCH ON CANADA
WATERLOO, ONTARIO, CANADA N2L 361
PHONE: (519) 885-1211 X3377
shortreed@civoffice.waterloo.ca

FYFE, WILLIAM S.
UNIVERSITY OF WESTERN ONTARIO
LONDON, ONTARIO, CANADA N6A 5B7
PHONE: (519) 661-3180
FAX: (519) 661-2179

USA

AELION, MARJORIE
UNIVERSITY OF SOUTH CAROLINA
DEPT. OF ENVIRONMENTAL HEALTH SCIENCES
COLUMBIA, S.C. 29208
PHONE: (803) 777-6994
FAX: (803) 777-3391
aelionm@sc.edu

ALLENBY, BRAD
AT&T AND LAWRENCE LIVERMORE LABORATORY
LLNL, P.O. BOX 808, L-1, LIVERMORE, CA 94551 USA
PHONE: (510) 423-7447
FAX: (510) 423-8746
allenby1@llnl.gov

BOZEMAN, BARRY
GEORGIA INSTITUTE OF TECHNOLOGY
SCHOOL OF PUBLIC POLICY
ATLANTA, GA 30332
PHONE: (404) 894-0093
FAX: (404) 894-0535
barry.bozeman@pubpolicy.gatech.edu

CICERO, PABLO	CALIFORNIA AIR RESOURCES BOARD UNIVERSITY OF CALIFORNIA, LOS ANGELES 9528 TELSTAR AV. EL MONTE, CA 91731 PHONE: (818) 350-6478 FAX: (818) 450-6108
CRANE, RANDALL	UNIVERSITY OF CALIFORNIA AT IRVINE DEPT. OF URBAN PLANNING IRVINE, CA 92717-5150 PHONE: (714) 824-7334 FAX: (714) 824-2056 rdcrane@uci.edu
DEJU, RAUL A.	DGL INTERNATIONAL/UNIV. OF CALIFORNIA, DAVIS P.O. BOX 3491 WALNUT CREEK, CA 94598 USA PHONE: (510) 945-6064 FAX: (510) 945-7457 rad3dglint@aol.com
DOÑEZ, FRANCISCO J.	GEORGIA INSTITUTE OF TECHNOLOGY 325741 GEORGIA TECH STATION ATLANTA, GA 30332-1075 fd12@prism.gatech.edu
ENGI, DENNIS	SANDIA NATIONAL LABORATORIES P.O. BOX 5800, MAIL STOP 1378 ALBUQUERQUE, NM 87185 PHONE: (505) 843-4122 FAX: (505) 843-4208 dengi@sandia.gov
FOLEY, CAROL	GEORGIA INSTITUTE OF TECHNOLOGY OFFICE OF THE VICE PROVOST FOR RESEARCH ATLANTA, GA 30332-0512 PHONE: (404) 894-3317 carol.foley@carnegie.gatech.edu
FRUMKIN, HOWARD	EMORY UNIVERSITY 1518 CLIFTON RD. ATLANTA, GA 30322 PHONE: (404) 727-3697 FAX: (404) 727-8744 frumkin@sph.emory.edu
MACARI, EMIR JOSE	GEORGIA INSTITUTE OF TECHNOLOGY GEOSYSTEMS ENGINEERING ATLANTA, GA 30332-0355 PHONE: (404) 894-1444 FAX: (404) 894-2281 emacarai@ce.gatech.edu

MULHOLLAND, JIM	GEORGIA INSTITUTE OF TECHNOLOGY ENVIRONMENTAL ENGINEERING ATLANTA, GA 303332-0512 PHONE: (404) 894-1695 FAX: (404) 894-8266 james.mulholland@ce.gatech.edu
PUMARADA, LUIS	UNIVERSITY OF PUERTO RICO - COHEMIS COHEMIS, MAYAGUEZ, PUERTO RICO 00681-5000 PHONE: (809) 265-6380 FAX: (809) 265-6340 cohemis-rum@rumac.upr.clu.edu
REPETTO, PEDRO	WOODWARD-CLYDE CONSULTANTS Director, Latin American Operations STANFORD PLACE 3, SUITE 600 4582 SOUTH ULSTER STREET DENVER, CO. 80237 PHONE: (303) 740-2667 FAX: (303) 740-2650 pcrepet0@wcc.com
RUSSELL, TED	GEORGIA INSTITUTE OF TECHNOLOGY ENVIRONMENTAL ENGINEERING ATLANTA, GA 30332-0512 PHONE: (404) 894-3079 FAX: (404) 894-8266 trusell@ce.gatech.edu
SANTAMARINA, J. CARLOS	GEORGIA INSTITUTE OF TECHNOLOGY GEOSYSTEMS ENGINEERING ATLANTA, GA 30332-0355 PHONE: (404) 894-7605 FAX: (404) 894-2281 carlos@ce.gatech.edu
SAUNDERS, F. MICHAEL	GEORGIA INSTITUTE OF TECHNOLOGY ENVIRONMENTAL ENGINEERING ATLANTA, GA 30332-0512 PHONE: (404) 894-7693 FAX: (404) 894-9724 michael.saunders@ce.gatech.edu
STRONG, JANET	OAK RIDGE NATIONAL LABORATORY P.O. BOX 2008, BUILDING 1505, MS6038 OAK RIDGE, TN 37880-6038 PHONE: (423) 576-0179 FAX: (423) 576-8646 pd9@ornl.gov

"Environmental Quality, Innovative Technologies, and Sustainable Economic Development - A NAFTA Perspective"

A Workshop Sponsored by

U.S. National Science Foundation (NSF)
U.S. Sandia National Laboratories (DOE)
Mexico's Consejo Nacional de Ciencia y Tecnología (CONACyT)
Canada's National Science and Engineering Research Council (NSERC)

Organized by

Georgia Institute of Technology
Center for Sustainable Technology (CST)

Universidad Nacional Autónoma de México
Instituto de Ingeniería/Instituto de Ecología

University of Western Ontario
Department of Civil Engineering

Breakout Groups and Group Leaders

Breakout Group 1
Track: Ground Water
Group Leader: F. Michael Saunders
Group Speaker: Juan Manuel Lesser

Breakout Group 2
Track: Air Pollution (Urban and Regional)
Group Leader: Pablo Cicero
Group Speaker: Rodolfo Lacy

Breakout Group 3
Track: Environmental Health Risk
Group Leader: Howard Frumkin
Group Speaker: Eduardo Palazuelos

Breakout Group 4
Track: Waste Water
Group Leader: Adalberto Noyola
Group Speaker: Enrique Mejía Maravilla

Breakout Group 5

Track: Solid Waste
Group Leader: Kerry Rowe
Group Speaker: Gabriel Auvinet and Margairta Eugenia Gutiérrez

Breakout Group 6

Track: Environmental Regulations
Group Leader: Carol Foley
Group Speaker: María del Carmen Carmona

Breakout Group 7

Track: Soil Remediation
Group Leader: Marj Aelion
Group Speaker: Alberto Jaime

Breakout Group 8

Track: Environmentally Conscious Manufacturing
Group Leader: Brad Allenby
Group Speaker: Alfonso Mercado

Breakout Group 9

Track: Technology Transfer and Property Rights
Group Leader: Barry Bozeman
Group Speaker: José Antonio Esteva Maraboto

"Environmental Quality, Innovative Technologies, and Sustainable Economic Development - A NAFTA Perspective"

BREAKOUT GROUP PARTICIPANTS

GROUP 1	GROUP 2	GROUP 3
Saunders	**Cicero**	**Frumkin**
(Leader)	(Leader)	(Leader)
Iturbe	Cruzado	Palazuelos
Lesser	Guzmán	Santos-Burgoa
Mazari	Lacy	Ostrosky
Crane	Revah	Allenby
Engi	Russell	Shortreed
Deju	Mulholland	Pumarada
Santamarina	Bozeman	Fyfe
Aelion		Foley
Repetto		Doñez
Rowe		Martin
Strong		
Downs		

GROUP 4	GROUP 5	GROUP 6
Noyola	**Rowe**	**Foley**
(Leader)	(Leader)	(Leader)
Mejía Maravilla	Gutiérrez	Carmona
Saunders	Auvinet	Giner de los Ríos
Crane	Rosales Aguilera	Bozeman
Deju	Sánchez Gómez	Engi
Aelion	Repetto	Russell
Fyfe	Macari	Cicero
Downs	Allenby	Shortreed
	Santamarina	Doñez
	Mulholland	Pumarada
	Fyfe	Martin

GROUP 7	GROUP 8	GROUP 9
Aelion	**Allenby**	**Bozeman**
(Leader)	(Leader)	(Leader)
Jaime	Mercado	Esteva Maraboto
Olivares	Olguín Palacios	Kuri-Parcuch
Saval Bohórquez	Russell	Mulholland
Santamarina	Engi	Cicero
Macari	Frumkin	Deju
Repetto	Shortreed	Doñez
Crane	Foley	Pumarada
Rowe	Martin	
Fyfe		
Strong		
Saunders		

"Environmental Quality, Innovative Technologies, and Sustainable Economic Development - A NAFTA Perspective"

Summaries from Breakout Discussion Groups

Breakout Group 1
Ground Water Contamination

Groundwater is a resource subject to controlled use and protection for regional and national areas. The coupled interaction of contaminated surface water sources and subsurface groundwater is frequently complicated by general shortages of water for potable use and ineffective waste disposal practices by industry, municipalities and individuals. This is a common occurrence in Mexico, U.S. border states, and industrial areas in the U.S. and Canada. Uncontrolled industrial development in concentrated areas has commonly resulted in highly contaminated and highly localized areas near industrial complexes in numerous North American locations. This high contamination, coupled with retardation of contaminant movement in soil systems, has resulted in the recognition of decade-old contaminant plumes only in recent years. These areas of contamination are common to North America, industrial areas of the globe, and are major opportunities for joint and collaborative application of technology and pursuit of research, testing and evaluation.

An opening presentation indicated that Mexican water resources are highly interconnected through the hydrologic cycle and that surface and subsurface contamination are strongly interactive. The concern for groundwater systems is frequently limited by immediate needs to address contamination of surface waters commonly relied upon for municipal and industrial water resources. A unique opportunity, furthermore, exists in the Mexico City Metropolitan Area with a population of 18-20 million. This region is a closed hydrologic system with groundwater recharge being routinely exceeded by water withdrawal such that aquifer levels routinely and consistency decrease at rates of fractional meters/year. The historical exploitation of a groundwater resource and its resulting negative impacts on the constructed infrastructure of the Mexico City area have resulted in contamination issues being secondary to more fundamental issues of meeting water needs, both potable and industrial. This situation, however, presents a fundamental opportunity to initiate study of numerous subsurface issues in the world's largest metropolis.

A brainstorming discussion resulted in the identification of areas of enhanced opportunity for research relative to groundwater systems. The opportunities are ubiquitous in North America and are focused on several fundamental areas. Contaminant identification, monitoring and fate pose complex issues. The role of heterogeneous soil matrices coupled with highly-variable levels of natural organic matter make monitoring of contaminants in subsurface environments difficult. The focus in the United States has been on assessing movement of aqueous-phase contaminants, with soil-phase contaminants assessed only indirectly unless found at neat levels, such as in non-aqueous phase liquids (NAPLs). This is further complicated in Mexico by a dynamically receding water table.

Contaminant speciation presents a major opportunity as well. The focus has been placed on identification of volatile, low-molecular weight, organic contaminants when opportunities exist for more complex petroleum-based and industrial-feedstock materials.

Coupled assessment of water supply and wastewater reclamation demands and investigation of water conservation and reuse are central issues in need of pursuit. The indication that the Mexico City aquifer is sinking at meter/decade rates is a singular indication of the need to assess these water source, use, supply and disposal issues in Mexico; depleted water resources in numerous North American areas present similar challenges.

Soils and sediments can be major reservoirs for contaminants in areas affected by high and uncontrolled industrial development. Fundamental understanding of factors affecting partitioning of contaminants between groundwater and soil and surface-water and sediments are central to assessing contaminant fate. In general, soil- and sediment-based contaminants can far outweigh the mass of contaminants in a river or aquifer section. However, control of soil and sediment contaminants through focused regulation is limited or nonexistent in North America. Given the transport of contaminants on soil and sediment particles to coastal estuaries and the interaction of contaminated groundwater with surface water through sediments, coastal sediments may present a most significant area of future collaboration among North American researchers.

Risk evaluation and assessment and its integration into decision making varies significantly throughout North America. Levels of environmental quality and economic status may appear to be in conflict in some instances. In others, risk assessment is considered to be a barrier to industrial development. A critical examination of the role of risk assessment and use in decision-making regarding sustainable economic development appears to be needed. This area is one in which considerable differences of opinion were voiced and appears to be a significant area for the integration of science, engineering applications, human and environmental health and industrial and governmental policy development.

Infrastructure issues are of major concern to Mexico with its aquifer subsidence issues and need to develop an expanded network of reliable water supply and reclamation systems. The U.S. and Canada are in situations where municipal and regional growth and expansion have significantly outgrown aging subsurface infrastructure and innovative approaches are needed. The extremes of these infrastructure needs could promote a focused opportunity to examine dual-water systems commonly used in some industrial countries (e.g., Japan). Advances in biological water reclamation systems, routinely applied to achieve advanced treatment levels, have many benefits when applied to systems where nutrient removal is of limited concern. Collaboration in these areas of wastewater reclamation need to be expanded due to extensive benefits and cost-savings associated with new technologies.

Cooperative Projects. In response to a workshop request, several priority issues were designated by the discussion group. These projects, in unranked order, were:

1. Workshop on Water Infrastructure in Mexico. Issues to be examined include Technical, Financial, Social and Technology and Environment Policy.
2. An education initiative focused on import of "environmental technology" into small and medium-sized firms and service enterprises.
3. Water Supply Infrastructure - Water supply piping in Mexico and U.S./Canada is a critical concern. A task force or research group should be initiated to identify problems; review cost-effective means of repair and replacement; system infrastructure education, repair and monitoring; and a workshop for consumers, users and technical experts.
4. Biodegradation and attenuation of contaminants was identified by the group as a major thrust and thereby a subgroup was formed. The report is included herewith.
5. Dams and reservoirs may be needed in Mexico to address water resource issues. The societal issues and environmental impacts of dam construction; cost-effective engineering techniques and practices; and rehabilitation of existing reservoirs, dams and surface-water quality need to be assessed.
6. Water Quality Issues in need of immediate pursuit with international teams include contaminant identification and fate; pollution prevention in watersheds; well-head protection; and combined sewer overflows.
7. Soil and Sediment Contamination. This area is key to elimination of groundwater and surface-water contamination since particulate-associated contaminants are the major sink for past contamination. While dissolved contaminants in waters are the immediate identifiable concern in most situations, the elimination of contamination in these media is critical to problem solution.

Breakout Group 2
Air Pollution at the Urban, Regional and Continental Scales

Most, if not all, large cities in North America share a common problem: air pollution. While Los Angeles used to be the most severely hit, the success of controls there and the great growth in Mexico City has led to the latter city now experiencing the highest levels of such photochemical oxidants as ozone. However, Los Angeles still has a severe problem, as do many other cities throughout the United States and Canada. Due to the interchange of not only air masses (e.g. along the border regions) but also fuels, technologies, products and automobiles (including trucks), it is prudent to consider developing strategies at the continental level to deal with the problems. This is not to say that the strategies should all be the same, but they should be coordinated. A number of issues in this area were addressed and ideas discussed.

Some achievements in air pollution control have been very noticeable in North America, including Mexico City. For example, carbon monoxide, sulfur dioxide and lead have been reduced over the last decades. Nevertheless, some problems have a very complex nature. In the case of ozone control, some local actions have been counterproductive at regional levels. Since ozone is formed in the atmosphere from the interaction of sunlight, nitrogen oxides and hydrocarbons, control strategies often are conflicting at local and regional levels.

Particulate matter, which was one of the triggering factors to initiate air pollution control poses new challenges. Its original control focused on point sources. Recently, it has been recognized the importance of the secondary component of fine particles. The fine secondary component of particles such as nitrates, re-suspension, as well as non specific sources have been more difficult to control.

In addition, control strategies have focused mainly on criteria pollutants, thereby relegating air toxics controls to local agencies capabilities. In the specific case of toxics, their full characterization may lead to a better understanding of the criteria pollutant sources and source strengths.

The working group identified ozone, particulate matter and air toxics as the most challenging atmospheric pollutants to control. The workshop was dominated by the gigantic challenge of Mexico City's air quality problem as it relates to mobile sources. Nevertheless, the general discussion of the table emphasized the relevance for North American issues. The topics of the discussion section were on diagnostic and modeling tools, the economic costs of air pollution and the potential economic controls as well as brief comments on new technologies.

The main questions addressed were the accuracy and applicability of the diagnostic and modeling tools and the need to identify their weak areas. Emerging concerns included the validity of the local atmospheric reactivity, meteorology, air quality, exposure, dose response, health effects, welfare costs, economic, productivity, economic, social implications, social, justice, and equity. Also, it was questioned the transportability and interpolability of current paradigms for cases like Mexico City.

In terms of economic costs of air pollution, on question was to how to value the costs of health effects, the value of a life, and in general contingency valuation. In the North American context very different idiosyncrasies are confronted , particularly how the valuation changes between countries. In addition to health effects it was questioned whether ecosystems should be evaluated from the point of view of the direct users or the more general recipients of such commodities that enjoy the intrinsic value of biodiversity. The proposal was to develop a study to assess the physical and economic issues, and how to value policy options, including uncertainties by the same team. One important emphasis of the study would focus on the process itself specifically on how different disciplines interact and the inclusion of equity and environmental justice in terms of recipients of costs and benefits of specific policies.

The discussion also focused on availability of new or alternative technologies including process modification such as the reformulation of gasolines, end of the pipe controls, continuous process monitoring, and retrofitting such as diesel ceramic filters. Additionally, the need for the diffusion and promotion of new innovative technology and the assessment of the full spectrum of effects of these alternative technologies were recognized. Examples included biotechnology, integrated transportation systems, as well as smart transportation systems.

Another important concern of the group was the lack of integration of the air pollution control process from characterization to the ultimate effects of the control policies in society. Very few projects have addressed the complexities of these problems with an integrated inter/multi-disciplinary team. Further, many policies have been preferred without consideration of the full range of impacts. It was agreed that future policy determinations should be based on life-cycle analysis of the different control options and technologies.

The group proposed an all-inclusive study for 2 to 3 cities in North America on how to design, implement and evaluate policies to control air pollution. The study should be addressing emission inventory and characterization, pollutant transport and transformation, air quality impacts, exposure assessment, analysis of dose response, health, welfare and vegetation impacts, economic costs and economic incentives of pollution control, equity of the proposed measurements and social justice.

The study may be complementary to the North American Research Strategy for Tropospheric Ozone (NARSTRO) proposal and should be available for this group in a cooperative environment. Areas that are not evidently considered by NARSTRO include the development of adequate tools for policy analysis for the technical, economic, and social aspects of pollution controls. Emphasis by this group should be focused on the best way to assess policy issues on a market and social bases.

To integrate a sound study proposal the group should seek seed funding to start the project development by proposing a meeting to identify the specific problems to address, the technical approach and the primary objectives of the study.

Breakout Group 3
Environmental Health Risk

The Roundtable on Environmental Health Risk began with a brief discussion of the issues by Dr. Palazuelos. He pointed out the difficulty of assessing risk, given scientific uncertainty and an intensely political environment, and the difficulty of balancing competing risks. The discussion was wide-ranging and animated. Participants reached agreement on several points, which can be divided into broad thematic issues, specific technical and programmatic needs, and proposed research projects.

Broad Thematic Issues

Each of the NAFTA countries has different environmental health problems, different economic realities, and different development priorities. Indeed, this is equally true within the NAFTA countries; the differences between Monterrey and Chiapas, or between Silicon Valley and South Bronx, are as profound as those between the U.S. and Mexico. Hence, the approach to environmental health problems must be flexible and based on local realities. The assessment and management of environmental health risks must be occur in the context of total risk. Each of the NAFTA countries, and each region, faces a different mix of risks, and may therefore evaluate and approach a specific risk differently.

In parts of Mexico, microbiological contamination of drinking water, and the associated risks of cholera, hepatitis A, and other infectious diseases, may eclipse the risk of chemical contaminants, while elsewhere chemical contaminants command a higher priority. Environmental health issues are multidisciplinary and multisectoral. Successfully addressing these issues requires the collaboration of engineers, epidemiologists, risk assessors, physicians, toxicologists, chemists, and others, representing industry, government, academia, and community and environmental groups.

Specific Technical and Programmatic Needs

The three NAFTA countries should move toward adopting common methods of measuring exposures and health outcomes. Exposures of interest include contaminants in air, water, food, and other media. Health outcomes include infant mortality, birth defects, cancer incidence, asthma incidence and prevalence, and other health statistics. Common methods of measurement will enable the countries to compare their experiences, identify problem areas consistently, and evaluate interventions.

The three NAFTA countries should move toward common approaches to risk communication and education. Specifically, there should be a NAFTA-wide commitment to the Right to Know about hazardous exposures in the workplace and the ambient environment, with free access to information in comprehensible form. Uniform practice throughout NAFTA would assure equity for all citizens, and would simplify the responsibilities of companies that operate across national borders. There should be effective public education about environmental health problems and their solutions, firmly grounded in science and with sensitivity to community needs. There should be avenues for public participation in debate and policy-making regarding environmental issues.

The three NAFTA countries should develop mechanisms to share knowledge about environmental health risks, their identification, assessment, and management, and about techniques of risk communication. Shared data bases, cooperative agreements involving government agencies, academic institutions, industries, and non-governmental organizations, should help diffuse knowledge and expertise throughout the NAFTA countries. NAFTA should be seen as an opportunity to build knowledge in all member countries.

The three NAFTA countries should cooperate to build human resource capacity in environmental health. In all three countries, and especially in Mexico, there are shortages of trained professionals in risk assessment, environmental engineering, industrial hygiene, epidemiology, and related fields. NAFTA should be seen as an opportunity to share in capacity-building in these essential fields.

Proposed Activities

Participants identified several possible projects that would be relevant, feasible, and informative. These include:

1. A project to identify criteria for risk assessment that can be used in all NAFTA countries, and that are compatible with NAFTA objectives for sustainable economic development and protection of the environment and health. The project would recognize differences among countries and expectations of the public in each country.
2. A study of risk communication needs in different countries and regions, taking into account differing levels of literacy, education, and expectations, different languages, and different availability of risk communicators.
3. A case study of a specific hazard, such as mercury, ozone, or hydrocarbons, in different cultures, climates, and economies of North America. This would enable comparison of the different NAFTA areas, in terms of the scientific and social issues that contribute to risk.
4. A survey of environmental health information that is currently available. What data bases are maintained, how consistent are they, how complete are they, and how accessible are they?
5. Develop an educational curriculum in environmental health, at one or more levels: elementary schools, secondary schools, undergraduate courses, and professional training. Consider the use of innovative methods such as long-distance learning, computer-based instruction, and case-based learning.
6. Develop communication tools to link environmental health professionals across the NAFTA countries, such as computer bulletin boards, trilingual newsletters, etc.
7. Develop internships or exchanges among the NAFTA countries for environmental impact and risk assessors, to enable them to observe each others' practices and share insights.

Breakout Group 4
Solid Waste

Following presentations by Gabriel Auvinet and Margarita Gutiérrez relating to the current situation with respect to solid waste in Mexico and the challenges facing Mexico, the Group focused on identifying areas requiring particular attention and where there would be considerable benefits from a coordinated effort by the NAFTA countries. Based on these discussions, the following five areas were identified as requiring a coordinated effort.

1. Regulation

Concern was expressed that existing regulations in Mexico, Canada and the U.S. with respect to solid waste management differ significantly and that this creates an unfair situation with respect to trade. The working group identified the following needs:

- To collect current regulations from Mexico, Canada and the U.S. (including U.S. EPA minimum guidelines and specific regulations relating to U.S. states bordering on either Mexico or Canada).
- Work towards a uniform method of classification of municipal, industrial and hazardous waste (waste shouldn't change its characteristics because it crosses a border).
- There is a need for a technical review by experts from the U.S., Mexico and Canada of current regulations with respect to:
    - Classification of waste
    - Evaluation of disposal requirements for different classes of waste
- Evaluation of potential long term impact on the environment resulting from the different regulations.
- How waste is managed (e.g. collection, handling, transportation and modes of disposal)
- How best to allow a "waiting" area where people in developing countries can obtain access to the post collection waste to recover salvageable materials. (ie. a post-collection re-cycling site).
- How regulations are enforced.

The following three individuals were identified as contact persons for the three countries: Gabriel Auvinet (Mexico), Pedro Repetto (U.S.A.) and Kerry Rowe (Canada). It was considered that this task could be undertaken by solid waste experts in the three countries relatively easily over a short time frame at modest cost.

2. Long Term Reliability of Waste Disposal Systems

It was recognized that although significant efforts are being made to reduce, reuse and recycle, and while alternative disposal methods (e.g. composting, incineration and pyrolysis) are gaining acceptance, there is still a need for landfilling and that the issue of long term reliability of landfill systems requires additional attention. Already there is considerable research being conducted in the three countries. However, there is a need for better communications and cross-fertilization of the research ideas. Areas requiring particular attention are identified to be as follows:

- long term performance of liner systems;
- clogging of leachate collection systems;
- seismic design and performance; and
- selection of appropriate design parameters.

The existing research suggests that there is a need to carefully compare clay liners and composite liners and that approaches such as that proposed under US EPA subtitle D may not be the best either technically of economically. More comparison is needed in the context of the hydrogeological conditions, the design and potential service life of leachate collection systems, the potential service life of geomembranes, recent research into the relatively high rates of diffusion of organic contaminants through plastic (geomembrane) liners, and recent research into the biodegradation of organic compounds in landfill leachate. This work item was ranked very high by the members of the Group and it was considered that considerable benefits could be gained for all countries by increased collaboration.

3. Research, Development, and Evaluation of Environmentally Friendly Alternatives to Landfilling

This would involve an examination of physical, thermochemical (e.g. pyrolysis and incineration) and biological (e.g. composting and biological raptors). There is already considerable research being conducted in Canada and the U.S. and significant interest in this research from Mexico.

This work item was seen to be a high priority item with considerable interest being expressed by researchers in the three countries. People interested in this area included Margarita Gutiérrez, Jim Mulholland and Susana Saval.

4. Socio-Economical-Cultural Issues and Waste Management Systems

This project would involve looking at the socio-economic and cultural differences between the three NAFTA countries and the implications of this for waste management. Issues that require consideration

range from the need for certain groups of individuals to derive their income by sorting through waste, recovering items that can be solid, the potential to produce useful products from waste, the potential to produce electricity from landfill gas and social, economic and regulatory restrictions on this. This was considered to be a medium priority topic.

5. Site Selection and Decision Making

This would involve an examination of the advantages and disadvantages of different approaches to site selection and decision making in the context of a risk evaluation. This was considered to be a medium priority item.

6. Monitoring and Contingencies

This would involve development of monitors that can provide feedback regarding upset conditions for air emissions whether they be from incineration, pyrolysis or landfill gas. There is also a need to develop improved monitors for groundwater and surface water monitoring which would improve the economics of monitoring waste disposal sites. This was considered to be a medium to low level priority item.

7. General Comments

There was a strong consensus among the Working Group that there was a need for improved communications between industry/academia and government in the three countries with respect to solid waste management. In particular, there is a need for improved communications by means of short courses, exchange programs and collaborative research. Courses should be coordinated by academic institutions in the countries. These courses could be extended to many Latin American countries.

Breakout Group 5
Legal and Regulatory Issues

María del Carmen Carmona presented an overview of the General Law of Ecology and other laws associated with natural resources in Mexico. She noted the importance of NAFTA and the Rio Summit in focusing attention on environmental issues and the need for significant policy reform. Many of her comments are reflected in the research needs identified below. Carol Foley noted, in her concept paper, that during the NAFTA debate organizations in both Canada and the U.S. conducted studies comparing environmental policies in Canada, the U.S., and Mexico. These studies found that each country had comparable environmental laws, with some containing provisions that the others lacked. The most significant difference between the environmental programs in the three nations is the relative institutionalization, strength, and experience of their enforcement agencies---Mexico created its enforcement agency (PROFEPA) in 1992. In spite of its relatively young enforcement program, Mexico has piloted an innovative regulatory approach that requires fewer highly skilled regulatory enforcement personnel. Mexico privatized enforcement audits in certain higher risk industries by requiring them to establish environmental management systems and pay third-party auditors to inspect them. It was clear from the resulting discussion that although Mexico faced daunting challenges in building its environmental policy infrastructure (human resources in technical and legal disciplines), that all three nations shared some common challenges.

Consensus-Based versus Legislative Approaches

Carmona noted that the bases for the legal systems among the NAFTA countries are different: in Mexico and Quebec, the basis is the Napoleonic Code, while in the U.S. and the rest of Canada, the basis is Common Law. The participants agreed that we have very little knowledge about the implications of this difference. To complicate matters, industry and government agencies are looking for alternatives to purely governmental systems of regulation such as covenants (Netherlands) and private-sector standards boards (ISO 14000).

Research is needed:

- To understand the legal and cultural dimensions of trade and environmental laws.
- To evaluate the effectiveness of legislative and consensus-based approaches.
- To develop and evaluate combinations of policies and programs (tools) that protect the environment and public health while allowing for flexibility in technological and managerial approaches in the regulated community.
- To assess the capacity of the NAFTA countries to implement traditional and alternative approaches.

- To understand the relationship between the various policy tools and innovation and technology choices in the regulated community.
- To increase the level of expertise in Mexico in environmental law, both Mexican and international.

Environmental Risk and Safety

In her presentation, Carmona noted that Mexico does not have a community right-to-know law and that without public participation, it will be difficult to incorporate new concepts like sustainability into policy. The U.S. and Canada have significant experience with worker and community right-to-know laws. The participants agreed that future efforts should include standardization of right-to-know data and processes for accessing it. They also noted that other standards, such as those for transporting hazardous materials, should be considered.

Research topics:

- Studies on options for standardizing right-to-know data, including consideration of what information the public needs to know.
- Studies on the effectiveness of right-to-know information transfer and dissemination programs and potential cultural dimensions.
- Studies on programs that enable the general public and industry to act in response to information generated under right-to-know laws (e.g., environmental justice, technical assistance programs, etc.).

Environmental Management Practices

The group discussed concerns about perceived differences in corporate environmental management practices across North America. One belief was that companies would relocate to Mexico because of perceived leniency in environmental regulation (making Mexico a "pollution haven"). Although several studies have been conducted by a variety of organizations (many of them in the border region), the participants noted that they were not aware of any effort to catalog or review them.

Research topics:

- Identify and review studies conducted on corporate environmental management practices in North America.
- Conduct baseline studies on corporate environmental management practices, stratified by country, by industry sector, and by size of company.

Breakout Groups 6 & 7
Remediation of Soils

The session on the remediation of soils was attended by university faculty from the Instituto de Ingeniería, UNAM, Georgia Institute of Technology, the University of South Carolina, the University of California, Irvine, the University of Western Ontario, Vice-President for Latin America from Woodward Clyde Associates, and representatives from PEMEX-Refinación. A keynote address was delivered by Dr. Alberto Jaime from the Gerencia de Protección Ambiental, Comisión Federal de Electricidad.

Soil Productivity from the Global Perspective

Although the topic area was on soil remediation from a NAFTA perspective, the importance of considering the topic from a global perspective was encouraged. One of the ways to improve world soil productivity (not associated with soil remediation of contaminated sites), was by the addition of inexpensive supplements to increase productivity. This simple measure could go a long way toward improving soil conditions globally.

Regulatory Considerations

Because of differences in environmental regulations and economic situations between Mexico, Canada and the United States, there are different incentives for implementing soil remediation. From the viewpoint of the PEMEX-Refinación, in the past, pollution prevention was top priority more than the remediation of contaminated soils. PEMEX-Refinación considered that the maintenance of existing tanks and facilities, reducing leakage of petroleum from piping, storage tanks and during fuel distribution, the installation of liners and the recycling of tank residues would have the greatest environmental benefit for the financial input. Containment of currently contaminated sites was considered to be the most practical manner to deal with soil contamination as opposed to more aggressive remediation strategies.

Mexican Experience with Soil Remediation

Several topics were discussed during the session and some debate was generated as to the priority of topic areas within remediation of soils. Several researchers in Mexico currently focus on soil remediation, but they reside primarily in 2 institutions, the Instituto de Ingeniería and the Instituto de Geografía. These researchers feel that they have the experience with hydrocarbons and Chromium (VI), respectively and soil remediation technologies for these contaminants, but have limited experience with other types of soil contaminants. Additionally, few field applications have been attempted and proven in Mexico. From the United States perspective, the group agreed that other technologies have been applied sufficiently to be termed proven technologies. These include containment, excavation and subsequent treatment, and in situ treatment or in situ bioremediation.

Several problems of particular concern to Mexico were expressed although some of these same situations exist in the United States and Canada. In Mexico, soil remediation is complicated by several factors including: sites contaminated by complex mixtures of inorganic and organic chemicals; contaminants which have been present at certain sites for long time periods; rivers and aquifers which are contaminated and therefore difficult to clean up; and particularly difficult sediment environments such as fine clays, swamps, and dredged materials abandoned on soils close to rivers. Additionally in Mexico there is a lack of experience in risk assessment and a lack of regulations for contaminated soils related to notification, action and clean-up levels.

The academic sector of Mexico has a great interest in participating and continuing research on soil remediation. However, there is some difficulty of access to contaminated sites, lack of economic support for research and development, lack of analytical infrastructure and limited interaction between industry and academic institutions. Other issues impeding the development of soil remediation technologies in Mexico include limited experience associated with field applications of soil remediation by Mexicans, few qualified Mexican personnel, and more contractors than well established companies carrying out the remediation efforts. Many of the companies performing soil remediation in Mexico are non-Mexican companies. Several Mexican participants expressed concern about the large number of companies selling technologies but not transferring the technology to Mexico; limited adaptation and innovation of remedial technologies to Mexico and Mexican problems; and that some of the technologies which have been attempted are not efficient at reducing contaminant levels.

Some sites in Mexico which may be appropriate for soil remediation projects because of the high level of contamination were identified including the Sta. Alejandrina swamp. The desired remedial technology for Mexico would be simplified technologies which are convenient, easily carried out and at low cost.

Future Needs in Soil Remediation

Several areas of research were suggested including the establishment of clean-up criteria for contaminated soil; the importance of adequate site characterization and assessment which includes the need to develop strategies for collecting information on the geologic setting of contaminated sites, the monitoring and evaluation of contamination; decision making for selecting a remedy with consideration of both present and future land use; calculating costs of the technology; understanding failure at the field level of attempted technologies; implementing technology transfer; and providing education and training for technical skills for soil remediation technologies. Also, it was stressed that all the above information and research must be carried out with an understanding of the regulatory issues and economic considerations which may be country-specific. In this light it was suggested that a document be written concerning soil remediation technologies to address issues and needs relevant to Mexico, Canada and the United States.

Breakout Group 8
Environmentally Conscious Design and Manufacturing

The session began with a presentation by Professor Mercado of El Colégio de México on his research regarding the environmentally conscious manufacturing activities of a Maquiladora plant owned by a Japanese parent. The standards set by the plant, as well as its management activities, were world class, and thus worthy of study as a model. The extent to which such desirable practices were widespread among similar Maquiladoras, especially those operated by smaller firms, was not clear. This introduction led into a spirited discussion of the value of environmentally conscious manufacturing techniques as a necessary component of any effort to reduce environmental impacts of economic activity over time, and the relationship between such technologies and their cultural and economic contexts.

Broad Thematic Issues

Evolution of environmentally and economically efficient technologies is critical throughout the NAFTA base. Such technologies will frequently not be "environmental technologies" as commonly thought of, but should considerably reduce the environmental footprint of manufacturing processes and products. Especially in areas where emission controls and remediation practices may not be robust, such technologies, which generate fewer and less toxic emissions than alternatives, may be even more important.

Environmentally and economically efficient technologies are culturally, geographically and demographically dependent. Like all technological systems, such technologies must be understood and evaluated within their economic, cultural and social context. What is appropriate in one area may not be appropriate, or even functional, under other circumstances. For example, a process that is used in small firms is likely to be heavily influenced by local context, while a process used by major transnationals is likely to become a global standard throughout that firm.

The life cycle perspective is critical. The environmental and economic efficiency of a process or product cannot be understood unless the relevant lifecycle considerations are included in any assessment. Methodologies and data for doing this, however, are primitive at best, especially outside the large transnational firms.

Technology diffusion, as opposed to technology development, is frequently the most difficult step in the commercialization of environmentally and economically efficient technologies. In order to make any difference in the environmental impacts of economic behavior, technology must be diffused throughout the economy.

This is a difficult problem, not well understood, and is heavily dependent on cultural context, management and firm behavior, and the incentives, implicit and explicit, which affect the relevant market sector.

There are a number of policy tools available to incent the evolution of environmentally and economically efficient technologies. These include command and control regulations, fiscal incentives, information generating tools such as eco-labels, international standards such as ISO 14000, requirements of non-NAFTA markets such as the European Union, and others.

Specific Technical and Programmatic Proposals

1. Neither the supply or demand side of technological evolution are well understood. Accordingly, studying the development and diffusion of environmentally appropriate technologies in the NAFTA region, perhaps through the use of comparative case studies, could be useful both specifically and as a contribution to the field as a whole.

2. It is important that both the environmental and technical aspects of potentially useful and environmentally preferable technologies be validated by a technologically sophisticated agency (usually not the environmental agency). This is being done for California, for example, by Lawrence Livermore National Laboratory. Accordingly, a NAFTA-wide technology validation program would be a useful public good.

3. The diffusion of environmentally preferable technologies frequently encounters barriers which are not obvious, and may involve subtle cultural factors. Case studies of such barriers in the NAFTA region would be a useful way to identify such barriers, and attempt to evolve means by which they can be minimized.

4. It is frequently the case that environmentally preferable technologies are not adopted because the economic benefits (which are usually positive, as environmentally preferable technologies are usually more efficient technologies) are not visible to managers. Accordingly, case studies based on rigorous cost-benefit analyses would be useful as ways to encourage the adoption of such technologies. Moreover, explanation of so-called "green accounting systems," and encouraging their implementation as part of adoption of "activity based costing" systems, is desirable.

5. While it is apparent that the definition of environmentally and economically appropriate technologies varies across geographic and cultural regions, the scope, magnitude and scale of these effects is not known. The NAFTA region is an obvious area within which to study these issues, possibly through the case study method.

Breakout Group 9
Technology Transfer

The workshop began with a presentation by Mr. José Esteva, Director, Centro de Innovación Tecnológica, UNAM. He observed that the cycle of transfer from the laboratory is quite long and that the laboratories are often separated from the institution producing the technology. This is particularly the case when university labs are the source of the scientific and technological results being transferred. Mr. Esteva noted that in Mexico universities play a particularly important role in science and technology production and the transfer of knowledge. In Mexico, it is often more economically rational to shut down a business, especially a small business, than to adopt innovative technologies. Small business needs technology suited for its particular context because it generally has no money for modification or retrofitting of technology.

Mr. Esteva described his own university-based engaged in technology transfer as a "virtual organization" with a small staff and a large network of labs, consultants and contractors drawn from companies. It is important to have linkage groups drawn not only from Mexico but North America. To the extent possible, these linkage groups should focus on clean technology as a source of business opportunity.

Mexico has the labs and scientific and technical expertise to support industry, but the network for technology transfer needs much development. There needs to be covenants with industry backed up with appropriate standards and certification required to promote technology transfer. Mr. Esteva provided a description of two different visions of technology transfer in Mexico, one "supply push" (university produces sciences and technology, then looks for users and developers), the other "demand pull" (industrial problems are taken to the universities). Both models are appropriate.

The group discussion followed from the talk. It was agreed that the focus should be on technology transfer issues affecting North America and interactions among nations. Some of the guidelines and recommendations flowing from the discussion included the following:

1. Provide a catalog of technological capabilities, including fully developed and pre-industrial technologies as well as sources of "know how."

2. Initiate trilateral workshops focusing on best practices for university-industry cooperation and on technology assessment and technological forecasting.

3. Use North American scientific and professional societies to promote technology transfer.

4. Recognize the cultural change inherent in technology transfer and the ways in which culture can inhibit technology transfer. This refers not only to the culture of national groups but also to the differences in institutional culture found among universities and industries.

5. Support the management of technology transfer, promoting cooperation in academic environments, tailoring transferred technologies to local needs and context.

6. Universities do not usually produce finished products, thus technology transfer should focus on collaborations that can be effective in the use of applied science and pre-commercial technologies.

7. Universities should employ "bridge" personnel who have extensive experience in both industries and universities and bring an understanding to the differences in norms and expectations.

8. Students can play a vital role in technology transfer through internships, placement in industry (while maintaining university contacts) and student projects.

The meeting concluded with an agreement to take the next step of proposing a technology assessment/technology forecasting conference focusing on NAFTA issues but inviting representatives from throughout North and South America.

BIOREMEDIATION WORKING GROUP

The role of Bioremediation in Environmental Restoration and Pollution Prevention

This section is a compilation of information gathered by Janet Strong during an unscheduled breakout session to further discuss issues of interest between the three NAFTA countries regarding Bioremediation. The meeting goals were to identify critical research needs, set priorities and approaches for the pursuit of these needs, and to identify individuals from the US, Mexico and Canada interested in collaborations.

A decline in the health of our environment not only reduces the quality of the water we drink, the fertility of the soil that produces the food we consume and the purity of the air we breathe but can also can lead to social instability. Thus, correcting and working towards environmental balance leads to socio-economic benefits within a city, state and nation. One of the most cost effective approaches for remediation of contaminated soil and groundwater involves *Bioremediation*, that is, the use of biological organisms (bacteria, protozoan, plants, etc.) for the degradation of contaminants (i.e. breaking them down into non-hazardous constituents). This technology is most successfully implemented at sites with relatively low contaminant levels or as a polishing step (residual remediation) once a point source or the majority of the contaminants have been removed by another process.

Bioremediation is an attractive, safe and efficient method of degrading hazardous compounds. Employing this technology within the subsurface (i.e. in situ) results in numerous benefits:

1) The contaminants are completely converted to carbon dioxide, water and inorganic compounds,
2) Natural populations of bacteria and plants can be used, thus minimizing the *current* concern over unintended adverse impacts from the release of genetically engineered microorganisms,
3) The technology does not generate a secondary waste stream that would require additional handling, storage, transportation or disposal,
4) The treatment cost is low, since there is no excavation and storage of soils
5) All organic degradation occurs below the surface. Contaminated water is not pumped above ground nor sludge excavated, thus there is minimal risk to workers and no reduction in air quality because volatilization of contaminants is minimized,
6) Integrating these in situ technologies can shorten the required treatment time,
7) Above ground impact can be minimized since there is no large scale excavation.

8) Infrastructure needs are less than with other treatment technologies, which makes bioremediation attractive for developing countries

Bioremediation can be either an active or passive process. Active bioremediation involves the addition of bacteria, plants, and nutrients into a contaminated region to accelerate the overall process of contaminant degradation. Passive remediation involves a minimal amount of intervention or disturbance and relies on the indigenous bacterial population and nutrient levels to degrade the contaminants. For passive remediation, activities primarily involve only contaminant monitoring. This latter process is typically slower than active bioremediation; however, it is very inexpensive and can be utilized where the environmental conditions limit the spreading of the contaminants to surrounding populations or groundwater regions.

The criteria for selecting active versus passive remediation may depend on the contaminant concentration as well as the *immediate* environmental and human health impact (see other sections). Given the fact that there are more contaminated sites than funds available for remediation, the most cost effective decisions must be made. In addition to funding limitations, an evaluation must be made to determine if a current technology exists to effectively treat the site.

If the decision is made to take an active role in remediation then the quickest and most efficient methods of restoring a hazardous waste site are to integrate several remediation processes. In situ biodegradation can involve the addition of nutrients, oxygen, electron donors, electron acceptors, surfactants, microorganisms, or all the above. These amendments can be introduced/coupled to a variety of other technologies such as permeability enhancements, chemical treatments, and/or physical treatments which will enhance the bioremediation processes. Modifying technologies from other industries such as hydrofracturing, deep soil mixing, horizontal drilling, etc. can accelerate in situ bioremediation by enhancing the distribution of bacteria or nutrients over a wider treatment zone.

Historically bioremediation has primarily focused on bacteriological processes for contaminant degradation. However, phytoremediation (the use of plants and the bacteria in the rhizosphere) has been successfully demonstrated to also degrade contaminants in areas where the contamination is limited to a relatively shallow zone. Phytoremediation has also been demonstrated using aquatic plants.

In addition to in situ contaminant degradation, processing industrial waste in a bioreactor is an efficient and cost effective option for pollution prevention. This involves degrading contaminants from industrial waste streams, on site as an "end-of-pipe" treatment. Information gained from laboratory based experiments and field demonstrations has resulted in the design of unique bioreactors that can efficiently degrade contaminants on-site and have the potential to significantly reduce environmental contamination. The energy input into these types of processes can be lower than chemical or physical based treatment processes. Because bioremediation largely uses native population of bacteria and plants, operation of these processes tends to occur without significant input of foreign energy sources thus, reducing adverse environmental impacts over the life cycle of the technology.

Pollution prevention permits a Proactive Strategy versus a Reactive Strategy. In addition to preventing waste generation, pollution prevention can also encompass recycling technologies which can be used to further offset operational costs. Furthermore, these activities can enhance public relations between the communities and industry.

Regardless of the methods used (active versus passive, bacteria versus plants, in situ versus bioreactors), successful bioremediation projects must be shared not only with the scientific and industrial communities but with the people living in that community. They are the real stake-holders. It is important to not only remediate the contaminated sites but also educate the community as to the processes involved, the cost, and the available technologies and the *limitations*.

As we examine the extent of environmental contamination from a global viewpoint, we cannot separate "there" from "here". It is obvious that contaminants do not remain localized within the boundaries of an industrial plant, within a city, within a country or within a nation. Thus, as we "share" these problems we can be most effective if we work together to solve them as well as preventing future problems. Our reactionary mode of *crisis management* is not effective, and as cities and countries become more populated the margin for error becomes smaller. Therefore, we cannot afford to ignore or postpone implementing current technologies and developing new solutions.

North American Bioremediation Working Group

USA Members
F. Michael Saunders
Marjorie Aelion
Janet M. Strong-Gunderson

Mexico Members
Adalberto Noyola
Susana Saval
Marisa Mazari
Eugenia Olguin
German Buitron
Francisco Guzman

Canada Members
Kerry R. Rowe

List of Position Papers (Alphabetical Order)

- **Aelion**, *Environmental Contamination and Site Remediation*
- **Allenby**, IEEE, *Sustainable Development and Industrial Ecology*
- **Bozeman**, *U.S. Federal Laboratories as Technology Transfer Partners with Mexico: Reflections on the Findings from "Industry Perspectives on Commercial Interactions with Federal Laboratories"*
- **Crane**, *Water and Waste at the U.S./Mexico Border: A PostNAFTA Perspective*
- **Deju**, *Sustainable Development and Environmental Conservation in the Americas*
- **Engi**, *Structuring a Collaboration between Mexico and the United States for managing Water Resources of mutual interest in the NAFTA context*
- **Frumkin**, *Healthy Environments and Environmental Health*
- **Fyfe**, *Earth System Science for Sustainable Development of Earth Resources*
- **Mulholland**, *Air Pollution Engineering: Source Reduction and Emission Control*
- **Pumarada,** *The Intergovenmental Meeting of Hemispheric Environmental Technical Experts 11/95 - Some of the Projects and Priority Issues*
- **Rowe**, *Waste Disposal: Some Items for Discussion*
- **Shortreed**, *Sustainability, Risk and Decisions*
- **Strong**, *Integration of Biotechnology in Remediation and Pollution Prevention Activities*
- **Downs,** *Water Supply and Wastewater Treatment/Reuse*
- **Macari,** *Geo-Environmental Concerns in North America*

ENVIRONMENTAL CONTAMINATION AND SITE REMEDIATION

C. Marjorie Aelion[*]

Introduction: Pollution Prevention and Remediation

Pollution prevention is the recent focus and the intent of the United States environmental and hazardous waste policy. Pollution prevention is the best solution to environmental contamination but the most difficult to achieve because it requires broad social and economic mandates from government and society. As with other issues of health, more efforts are given to "treatment" of problems than prevention. Even with environmental regulations in place today, voluntary and enforced compliance are often inadequate to prevent present and future contamination, and there remains for the foreseeable future a need to remediate environmental contamination resulting from past practices of contaminant disposal. Environmental restoration and remediation are realities for most countries, including the United States.

Despite our best intentions, site remediation, particularly that associated with the Comprehensive Environmental Response, Compensation, and Liability Act (CERCLA) or Superfund Act, is slow, ineffective and costly. Of the more than 1200 Superfund sites, few have been cleaned up and funds for the Superfund have not been collected since December 31, 1995 because of disagreement in the United States Congress on key issues including liability. Many of the dollars for remediation go to insurance company or corporate attorneys attempting to reduce the liability and costs for their clients. Regardless, this is part of the due process in the United States and is unlikely to change. Removing the liability from CERCLA is unlikely to increase the amount of dollars actually going to remediation, although it might reduce the amount of dollars spent in litigation. The larger question is, 'how can we improve this remediation process, and why is it so difficult?' This question

[*] Assistant Professor, Department of Environmental Health Sciences, University of South Carolina, Columbia, South Carolina 29208

raises issues of cleanup standards, particular difficulties associated with subsurface contamination, and levels of technology available for monitoring and remediation.

Cleanup Standards

Cleanup costs have been increased by regulatory criteria that often are based unrealistically on standards of uncontaminated areas, or drinking water quality. The need for expensive cleanups in areas which pose little risk to human health has been questioned (Lee 1993). What is the rationale of cleaning up ground water contamination to drinking water quality standards if it is unlikely that the ground water will be consumed for drinking water? This turns the focus of environmental cleanup to risk-based cleanup standards which occupies a regulatory gray area. Compound-specific cleanup standards provide a clear criteria for remediation. Risk-based cleanup standards leave the door open for interpretation. Risk assessment models are sensitive to assumptions, and incorporate a level of uncertainty particularly when based on limited data. Also, there are substantial individual and societal differences in the concept of acceptable risk. A survey of the opinions of the Canadian public versus those of Canadian toxicologists concluded that the general public had more negative attitudes and higher perceptions of risk toward the presence of chemicals in the environment (Slovic et al. 1995). Hence, risk-based assessments can be a double-edged sword, they are more realistic and cost-effective, but introduce more uncertainty and the possible rationale for "no remediation."

In addition to human health risks, there are uncertainties in environmental risks. The cumulative environmental effect of continued ground water contamination may be difficult to assess. Given that each incidence of ground water contamination is not an isolated event, the synergistic effect of multi-pollution is of great concern. Once an aquifer becomes contaminated beyond a certain point, it is unlikely that it can be restored to a previous uncontaminated state, and its use for drinking water will require extensive pretreatment.

Subsurface Contamination

One of the primary factors adding to the cost and difficulty of remediation of subsurface contamination is the environment itself. Surface water quality in the U.S. has improved over the past 30 years following efforts to cleanup polluted rivers which were routinely receiving industrial effluent. Although hydrophobic contaminants are persistent in surface water sediments, the water quality of rivers and lakes has improved. Some of the success associated with surface water remediation came not only from curtailing the inputs, but also because of the characteristics of surface water systems. Relative to ground water, transport and chemical reactions associated with surface waters are remarkably fast.

Ground water contamination presents a more difficult and long-term problem. Ground water is difficult to remediate because the subsurface environment is so complex. Physical, chemical and biological processes occurring in the gaseous unsaturated zone, the liquid saturated zone, and solid phase sediment make fine scale modeling of contaminant fate and transport difficult. Subsurface sediment characteristics including heterogeneities, such as small clay lenses or fractures, can impact both remediation technology and contaminant transport. Advances have been made in graphical representation, but basic data are necessary to validate sophisticated fate and transport models.

Development of New/Innovative Technologies

Environmental remediation technology has not changed drastically in the last 10 years. New technologies to remediate environmental contamination have been slow to develop and be tested in the environmental arena. We have tended to rely on proven technologies from other industries, such as wastewater treatment, for much of the biological treatment of contaminated ground water.

Combined technologies (e.g., combined air sparging, soil venting, and bioremediation) may be the most successful and versatile methods to treat a greater number of constituents and types of environments. In our system of environmental regulation, each environmental compartment, air or water for example, falls under different jurisdictions and regulations. Each remedial technology is tested individually, and yet the cumulative effect of combined physical and biological technologies may increase efficiency and effectiveness of remediation efforts.

While several technologies are continuing to be developed, at present most of the innovative technologies appear to have more success in above-ground reactors where operating parameters can be tightly controlled. The traditional technology of digging up contaminated sediment and incinerating or reburying it is no longer accepted, but is part of this ex situ treatment mentality. It is unlikely that removing all contaminated ground water or subsurface sediments to above-ground reactors will promote effective cleanup unless the contamination is localized. Hence, successful remediation may require in situ treatment technologies.

Unfortunately, the same properties that make modeling difficult make in situ treatments equally, if not more difficult. Bioremediation is often thought of as nonintrusive and cost-effective, and most concerns lie with the speed at which the process is occurring. The difficulty in implementation, transporting nutrients or electron acceptors to the bacteria and promoting rapid, sustained degradation over large areas is not trivial. Other innovative remedial technologies including in situ radio frequency heating for removal of sorbed contaminants, air sparging, and soil vacuum extraction have similar problems with implementation.

A new paradigm needs to be developed for in situ technologies which is not held hostage to the subsurface matrix but is applicable to different kinds of subsurface structures, sediments and levels of saturation. Many of the currently tested in situ treatment technologies rely on the introduction of physical structures such as monitoring wells into the subsurface. This type of system will always be limited by a "point" approach, and will have restrictions due to geologic formations. Intrinsic remediation (allowing nature to remediate naturally) is the ultimate in situ treatment technology. It will be less costly than active remediation and should be investigated as an alternative, but needs to be closely scrutinized to assure it is not used as an excuse to "do nothing."

Technology Transfer: Remediation and Monitoring

Technology transfer is critical and needs attention as we attempt to go to field applications of ground water remediation by small local groups. Many of the pollution generated begins with relatively small facilities, such as that generated from leaking underground storage tanks. Each state has different amounts of funds available for environmental remediation, and different technical expertise and economic incentives. The simplest, most easily employable technology may assure the greatest success. The more extensive remediation demonstration projects in the United States are often carried out using federal funds from the Department of Energy or the Environmental Protection Agency, and are on too large a scale for small states to implement. Development should be directed toward both technologically sophisticated and simple techniques. Larger, more costly and sophisticated remediation may be appropriate for large technologically driven commercial operations and governments. Smaller, less technologically driven private enterprises and state regulatory agencies with limited financial wherewithal and expertise may require simple technologies.

As with remediation technologies, the development of field sampling and monitoring instruments that are easy to use and inexpensive is important. A great deal of money is spent on sample collection. The current method of ground water sampling continues to be intrusive, and normally involves the installation of many monitoring wells. While the development of monitoring technologies such as direct push technologies for sampling ground water, soil and soil vapors limits surface disruption (Saunders 1995), it is still intrusive.

A great deal of money is spent for sample analysis for which a certified analytical laboratory is required. However, field monitoring instruments which can limit the number of certified analyses required at a site would limit costs associated with remediation efforts. Less precise measurements of contaminants from field monitoring equipment such as portable meters should provide sufficient information for contaminant plume identification and assessment. Limiting the cost of monitoring will go a long way toward limiting the cost of remediation.

Conclusion

How effective is any environmental policy and regulation? Regardless of all the regulations imposed in any location or country, the ultimate success lies in compliance and willingness to put forward some amount of economic investment toward improvement of existing facilities. In many cases the incentive for improvement has been economic. The threat of fines, or immediate cost savings due to process changes have driven pollution prevention. While this system is effective, additional incentive policies could serve to minimize the generation of contamination and future environmental pollution, and improve efforts toward environmental remediation.

Despite recognition that environmental remediation is important, there is no available technology that is clearly the most effective method for remediation of ground water and subsurface sediments (i.e., there is no "gold standard" of remediation technology), nor is there consensus on the best specific techniques for remediating specific contaminants or environments. Development of new technologies has not responded to environmental needs, perhaps in hopes that new regulations could exonerate or limit liability for cleanup. The current movement concerning ground water contamination which appears to be, "if no one is going to drink it, don't clean it up," is perhaps an admission of defeat. Technologies developed, used and refined for decades by other industries, also may prove to be applicable to in situ environmental remediation. More effort must be placed in developing new technologies to improve both monitoring and contaminant removal for the many and varied contaminants which pollute different environmental compartments and in combination threaten environmental quality and safe drinking water. Because the cost of remediating ground water to uncontaminated levels is not economically feasible for even the most affluent countries, it is obvious that present approaches are inadequate for the future.

References

Lee, R.T. Comprehensive Environmental Response, Compensation, and Liability Act. 1993 In Environmental Law Handbook 12th ed. Government Institutes, Inc. Rockville, MD.

Saunders, R.D. 1995. Direct push technology: Concept comes of age. Environmental Protection, 6(1):37-39.

Slovic, P., Malmfors, T., Krewski, D., Mertz, C.K., Neil, N., and Bartlett, S. 1995. Intuitive toxicology. II. Expert and lay judgements of chemical risks in Canada. Risk Analysis 15:661-675.

WHITE PAPER ON SUSTAINABLE DEVELOPMENT AND INDUSTRIAL ECOLOGY; IEEE ENVIRONMENT, HEALTH AND SAFETY COMMITTEE

Brad R. Allenby*

This White Paper on Sustainable Development and Industrial Ecology represents the views of the Environment, Health and Safety Committee ("EHSC") of the Institute of Electrical and Electronics Engineers, Inc. ("IEEE"). The IEEE is the world's largest technical professional society, with over 320,000 members in more than 150 countries. The EHSC was formed in July, 1992, to support the integration of environmental, health and safety considerations into electronics products and processes from design and manufacturing, to use, to recycling, refurbishing or disposal.

The EHSC believes that the original approach to environmental impacts and their mitigation, characterized by centralized "command-and-control" regulation targeted at emissions and existing waste sites, is far too limited to support the achievement of a sustainable economy. It must be replaced by a more comprehensive approach. As recognized in the document, Technology for a Sustainable Future, recently issued by the Office of Science and Technology Policy (OSTP), technology, science and environmental considerations must be integrated both in the U.S. and, eventually, throughout the global economy. This goal cannot be reached without developing a more sophisticated, coherent intellectual framework. This framework should support the development of the theories, methodologies, and data required by the technical community to achieve environmentally preferable processes, products, operations and technologies. Moreover, there is clearly a necessary role for government at all levels in supporting basic research in this area, particularly in developing the field of industrial ecology, and encouraging the development and diffusion of the resulting knowledge and technologies. Accordingly, we first propose here a model intellectual framework based on the field of industrial ecology. Using the U. S. Federal Government as a model, we then offer some brief suggestions on an appropriate role for government in helping to advance our understanding of industrial ecology. Believing that these issues extend beyond specific industrial sectors or areas of professional expertise, we have also chosen not to limit our comments to the

---

* Director of Energy and Environmental Systems, Lawrence Livermore National Labotarory

electrical or electronics area, although that may well turn out to be where the EHSC contributes most directly.

I. Industrial Ecology Conceptual Framework-

Sustainable development is traditionally defined as "development that meets the needs of the present without compromising the ability of future generations to meet their own needs".[1] It is worthy vision, but is inherently ambiguous, and somewhat value-laden as well, implying for some people, for example, redistribution of wealth or a need to restrict current consumption. Accordingly, while it provides a useful direction, it is almost impossible to operationalize. Standing alone, therefore, it cannot guide either technology
development or policy formulation.

Industrial ecology is the objective, multidisciplinary study of industrial and economic systems and their linkages with fundamental natural systems. It incorporates, among other things, research involving energy supply and use, new materials, new technologies and technological systems, basic sciences, economics, law, management, and social sciences. Although still in the development stage, it provides the theoretical scientific basis upon which understanding, and reasoned improvement, of current practices can be based. Oversimplifying somewhat, it can be thought of as "the science of sustainability."[2] It is important to emphasize that industrial ecology is an objective field of study based on existing scientific and technological disciplines, not a form of industrial policy or planning system.

The Design for Environment (DFE) infrastructure includes the legal, economic and other incentive systems, methodologies and tools, and data and information resources by which society provides the necessary and appropriate support for efforts by individuals and firms to implement the principles of industrial ecology. Examples might include the development of materials databases, based on industrial ecology R&D, which would provide simple, easily-accessed rankings of the environmental preferability of commodity materials in traditional uses.

Design for Environment (DFE) is the implementation of the principles of industrial ecology in the near term at the private firm or individual level. It may be broken down further into two separate sets of activities, at least at the firm level. Generic DFE includes the development of competencies, organizations, methodologies, and rules and tools across the firm which generally improve the firm's environmental performance regardless of specific design and production activities. Examples might include the development and deployment of "green accounting systems", "green business planning practices",
and "green specifications and standards." Specific DFE includes the development and deployment of rules, tools, and data sets intended to directly improve the environmental preferability of product and process design and operation. Examples might include development of product and process checklists, and DFE figure-of-

merit software to be included in CAD/CAM systems. In all cases, DFE activities require inclusion of life-cycle
considerations in the analytical process.

II. The Role of Government

There are essentially two roles for government to play in support of this industrial ecology framework (while we will refer for convenience to U. S. Federal Government entities in this section, the discussion is broadly applicable to international quasi-governmental organizations, other national governments, and state and local jurisdictions). The first, and perhaps most critical at this point, is supporting basic research in industrial ecology. The second is supporting the diffusion and implementation of technologies and practices based on industrial ecology, or development of the DFE infrastructure.

Basic research and development in industrial ecology is necessary to provide the objective understanding and support required for the integration of environmental considerations throughout the economy. It is also a necessary prerequisite for the development and implementation of economically and environmentally efficient regulatory structures, currently a critical policy deficiency. What this might entail may perhaps be best illustrated by a few examples, such as:

1. Planning and implementing a series of studies to understand and model stocks, flows, and logistics of material movements throughout both the US and global economies for all major industrial materials, including both renewables and nonrenewables. Environmental impacts and human/ecosystem exposure data could be mapped onto these models, providing the basis for developing environmentally preferable products and processes, and helping industrial sectors and labor markets adjust gracefully to an environmentally preferable world. Such knowledge, by the way, is also critical to support the development of valid, efficient, risk-based environmental regulations: indeed, it is difficult to see how environmental regulation can be effective in the long term without such data and models.

2. Developing integrated "industrial metabolism" models of energy production and use, linked where possible to technology, demographic and other systems, with risk assessment and technology option overlays. As in "1." above, this will facilitate the identification of optimal national and sectoral R&D and investment programs to produce environmentally and economically preferable (and, hopefully, eventually sustainable) energy, manufacturing, transportation, and other technology systems.

3. Developing integrated models of urban communities, including perhaps small relatively self-contained cities, larger cities with surrounding suburbs, and large megalopolises with decayed centers and most business activity decentralized throughout the suburbs. Such models would include transportation, physical infrastructure, food, energy and other systems. This would facilitate identification of

major sources of environmental impacts, patterns of activities which give rise to them, and potential environmentally preferable technological or mitigation options.

4. Developing integrated models of specific sectors of particular economic, environmental, or cultural importance - including, for example, the agriculture, forestry, extractive, electronic and automotive sectors - which could then be used to understand how they might be affected by an increasingly environmentally sensitive world. Such research could be particularly important in mitigating potential economic and employment shocks of discontinuous environmental, and/or related economic and regulatory, changes, and in supporting continued improvement in quality of life while reducing attendant environmental impacts.

5. One of the more robust hypotheses of industrial ecology to date is that rapid evolution of environmentally appropriate technological systems is a prerequisite for improvement of quality-of-life in an environmentally sensitive world. The fundamentals of technological evolution and diffusion throughout the economy are, however, poorly understood; still less do we know what optimum, or maximum, rates of technological evolution might be, what associated economic and labor costs and benefits might be (and how they could be optimized), and how such variables differ by class of technology. (For example, it is apparent that moving to a hydrogen-based energy economy will be significantly more difficult, and a far more lengthy process, than substituting for CFC-based cleaning systems in electronics manufacture.) Research into such issues may well produce valuable insights.

6. Investigating the interdependency of legal, economic, cultural, marketing, scientific and technological activities and policies as they affect environmental protection and the evolution of environmentally appropriate technological systems. Studies of different regulatory tools and approaches in terms of how private firm and consumer behavior subsequently shift, for example, could be quite useful in developing efficient private and public environmental management structures.

7. There are vast amounts of data available in buried, essentially inaccessible, files and databases across all levels of government, including both developed and developing nations. These data resources should be identified and prioritized, then made publicly available in a useful form. This can be linked in with a number of current initiatives; in the U. S., for example, the Federal Government should make this an initial goal of the National Information Infrastructure effort.

Such a research program should be supported in a number of ways. First, there are many existing programs and projects within the U. S. National Laboratories supported by the Department of Energy, NASA, the Department of Commerce, the Department of Defense, the Department of Transportation, the Department of the Interior, the Environmental Protection Agency and others which deal with aspects of industrial ecology, although they are not currently intended to do so, nor are they organized to take advantage of the knowledge they are generating. Examples include much of the energy research at the DOE National Renewable Energy Laboratory, Sandia National Laboratory, Lawrence Livermore National Laboratory, Lawrence Berkeley National Laboratory and elsewhere; the Partnership for a New Generation of Vehicle program involving the automobile sector and various national laboratories; new materials research at NIST, the DOE laboratories, and elsewhere; and much research throughout the government laboratory complex on pre-competitive environmentally conscious manufacturing technologies. (There are, of course, complementary R&D programs in academia and private industry, which must be brought into this collaborative research effort, but this White Paper focuses on the government role.)

Two actions with regard to these activities should be undertaken. First, a central government-wide point of contact, possibly in OSTP, should be established, not to control research or funding, but simply to assure institutional awareness of relevant government activities. As part of this institutionalization of industrial ecology, a survey of all pertinent projects should be undertaken to generate an "industry ecology project portfolio", and to encourage the integration of industrial ecology principles into such projects. Note that this action could generate significant data and benefits at little cost, as only existing, funded projects are involved.

Once this is done, research gaps and opportunities for intellectual cross-fertilization can be identified. Additional projects at the National Laboratories can be initiated if appropriate (and if funding is available). It is important, however, that the broader research community also be energized.

As a critical second thrust of such a research program, therefore, the National Science Foundation should establish a funding program for industrial ecology projects, possibly including the establishment of one or two Centers for Industrial Ecology at leading universities. This will generate momentum for industrial ecology research and teaching within the academic community, which is particularly critical where a field, such as industrial ecology, cuts across existing disciplinary lines and is therefore unlikely to be supported by traditional programs either within or external to the university.

The second general set of activities which must be undertaken by government involves the "DFE infrastructure" framework stage. Clearly, neither individuals nor private firms are able to develop on their own the overarching legal, regulatory and economic incentive structures which will be necessary to support the integration of environment into all economic activity. They are also not able to restructure existing regulatory systems - including environmental, but also including such apparently unrelated regulatory regimes as antitrust, consumer protection and government procurement - so that they avoid unnecessary interference with the achievement of environmental quality while still meeting their original purposes.

Similarly, some prioritization and reordering of environmental values, both among themselves (e.g., is Superfund, human carcinogenicity, or global climate change more important?) and in the broader context of other social values (e.g., employment, private property rights) can only be accomplished through the political process. While it is doubtful that an unambiguous, uncontentious prioritization of values is possible, some broader consensus is necessary to provide support for further progress: How, for example, can an engineer be expected to design a "green" product when what is environmentally preferable cannot be made clear? This will not be a trivial task. It will require, for example, the development of comprehensive risk assessment (CRA) methodologies, which evaluate and balance risks and possible benefits on a systems-wide basis. While such approaches have been suggested, no such methodologies yet exist, nor is it clear that the data or organizational structure necessary to support implementation of CRAs are currently available.

In a global economy where environmental perturbations are not restricted to political boundaries, it is obvious that such a domestic program, and others like it around the world, must be linked together in a collaborative international network. Existing international organizations, both quasi-governmental (e.g., UNEP and the OECD) and private (e.g., IEEE, ISO, and international public interest entities), must assume increased responsibility in this area. We also recommend that each national government appoint a high-level central contact point to facilitate the international exchange of data and establish collaborative programs for advancing the study and practice of industrial ecology. In keeping with the tenants of industrial ecology, this contact point should not merely represent the vested interests of environmental compliance and remediation, but should be technologically sophisticated and proactively interested in supporting the integration of science, technology and environment in all economic activities.

III. The IEEE EHSC Role.

As the world's largest international professional organization, and a forum where academia, government and industry meet in a neutral, professional environment, we believe that we bear a special responsibility to participate in the development and implementation of the nascent field of industrial ecology. This is particularly true because we believe the electrical, electronics and telecommunications sectors will be key in supporting the critical trend of providing increasing quality-of-life using less material and energy ("dematerialization" and "decarbonization"). We believe, in short, that we are enablers of industrial ecology and, eventually, the achievement of a sustainable economy.

Accordingly, we have already begun to play an important role as both a facilitator and a source of substantive expertise, as demonstrated, for example, by the IEEE EHSC annual International Symposia on Electronics and the Environment, begun in 1993 and now in its third year. This White Paper itself represents a desire to begin a process of public dialog on industrial ecology, as well as an effort to encourage necessary R&D activities and appropriate policy decisions in the public sector.

These activities should be expanded, however. We seek to make the IEEE EHSC an international resource of professional expertise in industrial ecology. As such, for example, our members could help peer review research proposals and programs, or identify data and methodological priorities for industrial ecology/DFE projects aimed at improving the economic/environmental efficiency of industry. Through articles written by our members (see footnote 2 for some existing examples) and other activities, we seek to further encourage the diffusion of industrial ecology and DFE practices and, more subtly, encourage the culture changes which can support the integration of science, technology and environment throughout the global economy. We do not undertake these activities simply because they are part of the high professional standards to which we adhere, although that is, of course, true: rather, we recognize them as our ethical obligation to our world and to our children.

IV. Conclusion

The IEEE EHSC believes that there are few more pressing public policy issues than the integration of environmental considerations into all economic activity with the ultimate goal of achieving a sustainable society. At the present time, however, our understanding of what this entails is preliminary and limited, in part due to the significant limitations of the existing intellectual framework within which environmental issues are defined. Accordingly, we propose a new framework within which to begin addressing these issues. Moreover, a necessary role for government at all levels in supporting research in the objective field of industrial ecology is identified, and some (mainly illustrative) suggestions made as to how that might be accomplished. Undoubtedly improvements to the above can - and should - be suggested. Nonetheless, we firmly believe that the time has come for all of us - whether in government, private industry, public interest groups, or academia - to begin to work together. We hope that the above framework provides a basis to begin to do so, and we pledge the IEEE EHSC to that course.

World Commission on Environment and Development (also known as the Brundtland Commission), Our Common Future (Oxford University Press, Oxford: 1987).

Further information on industrial ecology can be found in:

B.R.Allenby, "Industrial Ecology Gets Down to Earth," IEEE Circuits and Devices, January 1994, pp.24-28;

T. Graedel, B. R. Allenby and P. B. Linhart, "Implementing Industrial Ecology," IEEE Technology and Society Magazine, Spring 1993, pp.18-26;

B. R. Allenby and D. J. Richards, eds., The Greening of Industrial Ecosystems (National Academy Press, Washington DC: 1994);

E. Graedel and B. R. Allenby, Industrial Ecology (Prentice-Hall, Englewood Cliffs, NJ: 1995).

U.S. Federal Laboratories as Technology Transfer Partners with Mexico: Reflections on the Findings from "Industry Perspectives on Commercial Interactions with Federal Laboratories"

Barry Bozeman*

Based on findings from a recent study of the U.S. federal laboratories' roles in technology transfer and industrial development within the U.S., this paper speculates about the likely success of collaborations between U.S. federal laboratories and Mexican industry. The recent National Science Foundation-sponsored study, Industry Perspectives on Commercial Interactions with Federal Laboratories (Bozeman, Papadakis and Coker, 1995) found evidence from a study of 239 federal laboratory-industry technical interactions, that the average net (costs subtracted from benefits) economic benefit from the projects is in excess of $1,000,000 but that there is considerable variance in net benefit. The purpose of this paper is to review the chief findings of the earlier study and suggest possible lessons for effective collaboration between U.S. federal laboratories and Mexican business.

The focus of the earlier study was on U.S. firms' commercially-relevant technical interactions with federal laboratories. The report presented data from questionnaires mailed to industrial organizations which have interacted with federal laboratories during the past five years. The questionnaires asked about experiences with particular projects (identified by project title and summary) with particular federal labs. The respondents were employees of those firms, generally individuals designated as the directors or administrative heads of the firm-federal laboratory project in question. The data include 229 federal laboratory-industry interactions from 27 federal government laboratories, including most of the leaders in federal laboratory-industry commercial activity.

In most instances, the results were quite favorable. For the set of interactions examined in the study (Bozeman, Papadakis and Coker, 1995), 22% had already led to a new product, process or service being marketed- a high rate given the fact that most of the projects began after 1990. This rate compares favorably to productivity rates from R&D performed exclusively by firms.

Comparing those companies who had already developed projects to all other companies, the ones with products developed have the following characteristics: (1) smaller than the average for all companies in the data base (12,000 employees on

* School of Public Policy, Georgia Institute of Technology, Atlanta, Georgia 30332

average vs. 25,000 for the sample), (2) high levels R&D intensity (R&D employees as a percentage of total employees), (3) established more recently (average 27 years, compared to 45 years for all firms in the database).

The study showed that the industrial partners were "satisfied customers." More than 89% of respondents agreed that the interaction was a good use of their company's resources. Even many (41%) who reported that costs exceeded benefits nonetheless have a high level of "customer satisfaction." Job creation was the single criterion by which the laboratory-industry interactions could not be said to have been particularly successful. It is important to note, however, that the study focused only on jobs created by the company (not by vendors or suppliers) as a direct result of the project. Using this relatively conservative approach, more than 90% of the projects did not result in a single new hire.

The measured cost-benefit was generally favorable but required considerable care in interpretation. In general, the lab-industry interactions are, in aggregate, a "good investment." But this does not mean that every interaction pays off handsomely for the firm. There are vast differences in outcomes, with some firms having huge pay-offs and others suffering substantial losses. The average level of benefit reported is $1,548,03. The average level of cost is $448,765. However, many firms report either no cost (43%) or no benefit (17%). Focusing only on the firms who sustaining both costs and benefits, the average benefit is $1.8 million and the average cost is $544,000, a return of more than three to one.

One of the more straightforward ways to measure net benefit is to simply subtract costs for benefit to determine net benefit. Doing so shows that the average (arithmetic mean) net benefit for all projects is $1,087,500. The average net benefit number is highly skewed by a few "big winner" projects. Overall, projects are somewhat more likely (49%) to have net benefits than net zero (18%) or net costs (33%).

One of the strategic questions suggested by the results is whether to develop policies to help grow small business or whether to pursue "big winners." Just a few of the projects have such enormous benefit (in excess of $10,000,000) as to provide an excellent justification for lab technology commercialization policies. But there are good arguments for the labs serving as a spur to small business (as many do).

Another important strategic issue is the role of basic research in commercialization. Product development-focused activities are likely to provide substantial benefits. Basic research does not, on average, enhance the likelihood of some commercial pay-off from interactions. But those projects that are "big winners," the ones with huge returns, are disproportionately basic research projects. This seems to indicate that just as basic research itself is a high risk/high pay-off activity, basic research aimed at commercialization seems to have the same character.

Finally, a recent study (Bozeman, 1996) comparing DOE multi-program national laboratories to all other laboratories in the data base, found that the DOE labs were tended to be somewhat more successful in their commercialization and technology transfer efforts. This is particularly significant for present purpose in that some of these DOE laboratories (i.e. Sandia National Laboratories, Los Alamos National Laboratory) are in close proximity to Mexico.

Implications for Mexican Industry Partnerships with U.S. Federal Labs

It is not clear just how many lessons of the U.S. study might be generalizable to case of technical interactions between Mexican industry and U.S. federal laboratories. But this suggestion provides some extrapolation.

1. Familiarity

One factor related to the success of the U.S. projects was familiarity between the personnel at the federal labs and the industrial partner. Some of the most successful relations had transpired over many years of informal interaction. This implies that U.S. laboratory-Mexican business partnerships might not pay-off as quickly as one might hope, that time would need to be invested in "getting to know one another."

2. Job Creation

U.S. laboratory-Mexican business partnerships should not be built solely on the promise of job creation, this would likely lead to considerable disappointment. The value of the interactions seems to be development of human capital at the companies and develop of new commercial products (at the lab or at the company) rather than new jobs. New jobs are likely to come from creation of whole new enterprises rather than additional jobs at partner companies.

3. Big Projects vs. Small Projects

The impact of the typical interaction is not great. The reason the net benefit is in excess of $1,000,000 is that a few projects skew the results. A strategic decision needs to be made. Is there a stronger interest in aggregate economic development? If so, projects should be cultivated that involve large, technically-intensive business, possibly in a consortium relationship. Is there more interest in regionally-dispersed economic development and spurring small business? If so, the strategy suggests a more distributed approach, but great care must still be taken to match up the most appropriate companies with the most appropriate lab technical personnel. Most small business-federal laboratory projects in the U.S. do not exhibit great benefit. Perhaps one reason is the greater value ascribed to basic research and pre-commercial R&D, both more easily appropriated by companies with more technical resources.

AIR POLLUTION TRENDS AND CONTROLS IN NORTH AMERICA: LOS ANGELES, MEXICO CITY, AND TORONTO

Pablo Cicero-Fernández*

INTRODUCTION

Air pollution levels in urban areas in North American urban areas have become an important issue of political agendas. While in some cities of the United States and Canada air pollution controls have been implemented for at least 30 years, in the Latin American countries these controls have appeared only recently. Here, a comparison between Los Angeles, Mexico City and Toronto is presented.

The cities were selected based on their population, stage of development, known air pollution problems, and because they represent different spectrums of controls in the three different countries belonging to the North American Free Trade Agreement. Another important consideration was the existence for more than a decade of operational air pollution monitoring networks, thus ensuring availability of reliable data. To compare the air quality of the three metropolitan areas, the Canadian, Mexican and the United States standards, as well as the World Health Organization guidelines were used. The structure of environmental agencies and the implementation of pollution control programs is also explored.

AIR QUALITY STANDARDS AND CRITERIA

To compare the air quality of the three metropolitan areas, the maximum pollutant levels of selected stations or zones were used on hourly, 8 hour, 24 hour, three month and yearly basis when pertinent. The comparison was performed in accordance with Canadian, Mexican and United States standards and World Health Organization (WHO) recommended guideline values. Table I shows how alike are the standards adopted in each nation. In general, the WHO levels are the most stringent. These recommended levels are based on toxicological and epidemiological data, although the final regulations often obey political forces, thus resulting in the marginally different standards.

AIR POLLUTION IN LOS ANGELES, MEXICO CITY, AND TORONTO

A demographic and emissions summary for the three cities are presented in Table II. Table III show the recent air quality maximum values of the highest air monitoring station in the study areas.

---

* Environmental Science and Engineering Program, University of California, Los Angeles, CA 90095-1772

LOS ANGELES

Located at a latitude of 34° N, Los Angeles has a rather benign Mediterranean climate compared to other places in the United States. The sunny skies of Southern California, its topography, and its geographical location are factors very important to determining the air pollution problem of Los Angeles. During the summer days, solar radiation reaches its maximum, and a typical "sea breeze pattern" promotes transport of air masses west to east. The surrounding mountains at the north and east promote stagnation of the air at the east end of the basin. The influence of source areas in the west impact receptor areas to the east. Damage to vegetation has been observed in the San Bernardino mountains 100 km from the main emission sources.

The metropolitan area of Greater Los Angeles is known for air pollution regulatory purposes as the South Coast Air Basin (SoCAB). This area is under the jurisdiction of the South Coast Air Quality Management District that includes all of Los Angeles and Orange counties and portions of Ventura, San Bernardino and Riverside counties. In the past the Los Angeles Air Pollution Control District, created in 1947, enforced air quality regulations well before any other air pollution control agency in the United States. The SoCAB has an area of 17,087 km^2 and a population of over 12 million inhabitants, being one of the most diverse urban areas in the continent. In terms of transportation, 8 million vehicles are registered in the SoCAB. Only less than 20% of the population in this basin use public transportation. The total emissions in the basin are estimated to be 3.2 million tons per year. Mobile sources are estimated to contribute half of the hydrocarbons, three quarters of the nitrogen oxides, and almost all the carbon monoxide emitted in the Basin.

The Los Angeles photochemical smog problem is well known, although the situation has been improving from a highest concentration 0.68 measured in 1955 to 0.30 parts per million (ppm) in 1992. Despite this dramatic improvement, the concentration of ozone (O_3), as well as nitrogen dioxide (NO_2), carbon monoxide (CO), and fine particles (PM_{10}), in Los Angeles are the highest in the United States.

The achievements in Metropolitan Los Angeles has been due to increasingly stringent air pollution controls, even with increasing population growth over the past 30 years. Nevertheless, to achieve air quality goals in the near future, the controls will affect almost all the residents of the area in a direct way. Electric cars, water-based paints, and the ban of barbecue lighter fluid are just a few of the measures included in the South Coast Air Quality Plan approved in 1991. The leadership in air pollution control by the local agency, the South Coast Air Quality Management District, along with the California Air Resources Board has had an enormous impact on other states concerning "ways to cope with pollution" and in many cases has helped to develop state of the art technology used to set federal environmental policy by the U.S. Environmental Protection Agency.

MEXICO CITY

Mexico City is situated in a valley surrounded by mountains. It is located at an average altitude of 2240 meters above sea level on a site which was previously a series of lakes, in an area of 7860 km^2, at a latitude of 19.5° N. Compared to Los Angeles and Toronto, it is the closest to the equator. Also due to its altitude, it is the most susceptible to receive solar radiation of high energy throughout the year. The climatic conditions, which in winter include an absence of winds and stationary masses of cold air, as well as surrounding mountains result in conditions which favors thermal inversions. The frequency of temperature inversion is very high during the winter months, although they usually break by noon. Mexico City has a mild climate with a rainy season from June to September and a dry season the rest of the year. The dominant wind blown from the north although micro-scale winds dominate in the early morning due to a valley-mountain effect, in particular in the western and southern areas. The dry lake of Texcoco, to the east of Mexico City, has an incipient vegetation cover. To the west and south are high mountains, with an average altitude higher than 3000 meters. Mexico City has a population of 18,000,000, and a fleet of has 2,500,000 vehicles. In contrast with Los Angeles, on the average one vehicle serves 7.2 inhabitants, and the use of public transportation is close to 80%. Direct man made emissions are estimated to contribute 3,700,000 tons per year, and including erosion and other natural degradation processes the total emission estimate climbs to 4,400,000 tons per year. Mobile sources dominate man-made emissions of hydrocarbons, carbon monoxide and nitrogen oxides. Natural sources play an important role in the generation of particles mainly from erosion, as well as biogenic hydrocarbon emissions.

In terms of air quality, the most publicized problem has been photochemical smog, which has been increasing steadily. Between 1986 and 1995, average non-attainment of the Mexican O_3 standard increased from 30% to 76% of the days. This situation exacerbated during March, 1992, when ozone levels reached 0.47 ppm. A less publicized problem is particles concentrations, both of total suspended particles (TSP) and its fine fraction (PM_{10}). Total suspended particles reached a concentration of 1494 mg/m^3 and fine particles reached 607 mg/m^3 in 1989. CO average maximum values of 31 ppm for 8 hours have been observed in 1987 but have been decreasing, reaching 22 ppm in 1994. SO_2 average daily maximums have shown decreasing levels from 0.45 ppm in 1987 to 0.22 ppm in 1994, due to a reduction in the sulphur content of fuels. A decreasing trend also has been observed for lead, from an annual average of 3.6 mg/m^3 during 1981 to 0.4 mg/m^3 in 1993.

Mexico City has been slowly moving to clean up the air. In the past, the air pollution arena has been subject to bitter encounters between federal agencies and federally owned industries. For example, the former Secretariat of Urban Development and Ecology, the Federal District and PEMEX have been involved in uncoordinated efforts since 1986, including the reduction of sulfur emissions and the reduction of lead in gasoline. Recently, a “new” policy approach in Mexico has initiated another stage of ample cooperation between these sectors. Later in this paper, a detailed description of Mexican air quality management will be presented.

TORONTO

Greater Toronto is located on the eastern shore of Lake Ontario and differs dramatically with Mexico City and Los Angeles not having some physical topographic constrains. Greater Toronto comprises the cities of Etobicoke, Scarbourough, North York, East York, York and Toronto as well as the regions of Peel, Halton and Durham. Toronto is at a latitude of 43.6^o N and experience continental climate with a temperate summer. It is the most populated metropolitan area in Canada with 3,900,000 people and over 3,800,000 vehicles. By extension, Metropolitan Toronto has an area of 6220 km^2 with a population density of 627 inhabitants per square kilometers. Toronto's emission inventory of 1,700,000 tons per year, is less than half of Mexico City's emissions. Three quarters of the CO emissions, two thirds of NO_x emissions and about a third of the hydrocarbons emissions are produced by motor vehicles. Its geographical location promotes mid- and long-range transport as a source and as a receptor. The transport come from nearby Canadian and United States cities.

Photochemical smog is present in Toronto and most of the Windsor-Quebec corridor during the summer. Its location provides insulation starting before 5 am during the summer. Toronto has experienced O_3 concentrations of 0.163 ppm (maximum of the average of 3 high years), exceeding the 0.082 ppm Canadian standard 24 days per year. Particulate matter also has being a problem, with 24 hours concentrations of TSP as high as 720 mg/m^3 during 1991, but the annual average concentrations have been dropping steadily since 1974. In 1991, Toronto's annual mean was 67 mg/m^3. Lead presented 3 exceedances of the Canadian standard in 1991, although experienced a steady decrease in the past decade. Toronto also has registered a decrease in its levels of CO from a maximum 8 hour average of 18.4 ppm during 1984 to a maximum of 3.6 ppm in 1991 even with a consistent increase in vehicle kilometers traveled in the past 10 years. In the case of SO_2, there has been a steady decrease in ambient concentrations.

In addition to the common criteria pollutants the Provincial Government of Ontario has implemented regulation for heavy metals such as Cr, Mn, Ni and Cu. Toronto's success in controlling air pollution is mostly due to the cooperation of Environment Canada and the Ministry of the Environment of the Provincial Government of Ontario. Many programs are supported at the municipality and local level through transportation management plans. In terms of industrial emissions, new and retrofitting programs are provided under the Inter-Ministerial Management Plan for NO_x and Volatile Organic Compounds. In addition to the important health effects issues, atmospheric air pollution controls are driven by regional and global concerns such as acid rain, regional ozone, and global warming. These concerns focus on forest, aquatic and agricultural ecosystem protection in contrast to the local emphasis in Los Angeles and Mexico City.

THE MEXICO CITY PROCESS

The achievements in Metropolitan Los Angeles have been due to continuous decisive action in controlling air pollution, even with the population growth of the past 30 years. Toronto's success in controlling air pollution is similar due to local, provincial and federal government regulatory effects.

Mexican officials have been learning from the Los Angeles and the Toronto approaches. But in Mexico City, air pollution control history has been different. Being a Federal District, the local government is effectively the federal government. The first institution to deal with the problem was the Ministry of Health starting in 1971. In 1982, to overcome the lack of coordination among agencies, a ministry (Secretariat of Urban Development and Ecology) was formed to deal with urban and ecological problems. More recently it was reorganized as the Secretariat of Environment, Natural Resources and Fisheries. In 1989, an amendment to the environmental legislation delegated jurisdiction to cope with air pollution to the local government, the Federal District Department (DDF). More recently a Metropolitan Environmental Commission including the neighboring State of Mexico started to play a role very much like the South Coast Air Quality Management District for the Metropolitan Area of Mexico City, although the new commission also addresses ecological problems other than air pollution.

Specific actions to reduce emissions of air pollutants in the Mexico City Metropolitan Zone include: the substitution of natural gas for heavy fuel oil in the two power plants in the metropolitan area (1986); a day without your car (1989); the implementation of a mandatory motor vehicle inspection and maintenance program

(1989); the use of oxygenated gasolines (1989); unleaded gasoline introduced an a large scale (1990); diesel with reduced sulfur content is mandatory for sale in the zone (1990); taxi and micro-bus substitution program (1991); closing of a major refinery (1991); restrictions on the use of industrial fuels with high sulphur content and promotion of gas oil (1991); and the adoption of emission standards for new vehicles that requires the use of three way catalytic converters (1991). Additional programs initiated in 1992 included a gas retrofit program to passenger and cargo vehicles, and contingency measures focused on the industry that also includes retrofitting.

In the search for off-the-shelf actions, the previously mentioned "a day without a car" program was implemented. This measure supposedly reduced by 20% the private vehicles on the road. Potential problems with this program include an increase in driving hours or the procurement of older cars (more polluting) from outside of the Basin. The media has been playing a prime role by pressuring the authorities. Also public participation has increased in the ecological arena and may be viewed as a part of the development of democratic aperture in Mexico in the recent years. International cooperation from Germany, United States, Japan, and the Province of Quebec, has helped Mexico City in developing diagnostic, forecasting, and modeling tools. Areas that have not been fully explored in the case of Mexico City include the impact of mid- and long-range transport of pollutants generated in the Basin, and receptor modeling to help validate the current inventory and present control policies. Emission inventory development has to be refined, including biogenic emissions. Other areas open to research includes the identification and control of toxic air contaminants. Finally economic instruments to develop flexible and effective air emission control policies may be explored, such as emission trading and other economic incentives.

ACKNOWLEDGEMENTS

I would like to thank Dr. Andrew J. Forester from Dartside Consulting, Toronto, Ontario, David Niemi from Environment Canada, Hull, Quebec and Sergio Sanchez from DDF, Mexico City, D.F. for providing information and Michael Benjamin from the California Air Resources Board, El Monte, California for his useful comments.

BIBLIOGRAPHY

California Air Resources Board, California Air Quality Data, Summary of 1992 Air Quality Data. Sacramento, 1993.

Comisión Metropolitana para la Prevención y Control de la Contaminación Ambiental. La Contaminación Atmosférica en el Valle de México, Acciones para su Control 1988-1994. Mexico City, 1994.

Departamento del Distrito Federal. Integrated Program to Combat Atmospheric Pollution in the MCMA Emergency Program. Mexico City, Departamento del Distrito Federal. 1990.

Government of Canada. The State of Canada's Environment. Ottawa, Canada Communication Group, 1991.

Lloyd, Alan C.; Lents, James M.; Green, Carollyn; Nemeth, Patricia. "Air Quality Management in Los Angeles: Perspectives on past and Future Emission Control Strategies" JAPCA. 39, 5, 696-703, 1989.

Ontario Ministry of Environment, Air Quality in Ontario 1991. Air Quality and Meteorology Section, Air Resources Branch, 1992.

World Health Organization (WHO). Air quality guidelines for Europe. Copenhagen, WHO regional Publications, European series No. 23, 1987.

TABLE I. AIR QUALITY STANDARDS AND GUIDELINES

Parameter	Canada	Mexico	U.S.A	W.H.O.
CO				
1 hour (ppm)	31		35	35
8 hours (ppm)	13	11	9	9
Pb				
24 hours (mg/m^3)	5			
30 days (mg/m^3)	2			
3 months (mg/m^3)		1.5	1.5	
annual arithmetic mean (mg/m^3)				0.5-1.0
NO_2				
1 hour (ppm)	0.213	0.21		0.21
24 hours (ppm)	0.106			0.08
annual arithmetic mean (ppm)	0.053		0.053	
SO_2				
1 hour (ppm)	0.344			0.12
24 hours (ppm)	0.115	0.13	0.14	0.017-0.044
annual arithmetic mean (ppm)	0.023		0.03	0.015-0.023
O_3				
1 hour (ppm)	0.082	0.11	0.12	0.075-0.100
8 hours (ppm)				0.050-0.060
annual arithmetic mean (ppm)	0.015			
PM_{10}				
24 hours (mg/m^3)		150	150	70
annual arithmetic mean (mg/m^3)		50	50	
TSP				
24 hours (mg/m^3)	120	260	260	150-230
annual geometric mean (mg/m^3)	60	75	75	60-90

concentration
(year)

TABLE II. A BRIEF COMPARISON

Parameter	Los Angeles	Mexico City	Toronto
Population	12,000,000	18,000,000	3,900,000
Vehicles	8,000,000	2,500,000	3,800,000
Total man-made emiss. (ton/year)	3,200,000	3,700,000	1,700,000
Area (km^2)	17,087	7,860	6220
Pop. density (hab/km^2)	702	2290	627
Veh. density (veh/km^2)	468	318	611
Emiss. density (ton/year/km^2)	187	471	273
Emiss. per capita (kg/year/hab)	267	206	436
hab. per vehicle	1.5	7.2	1.0

concentration
(year)

TABLE III. AIR QUALITY IN THE SELECTED CITIES

Parameter	Los Angeles	Mexico City	Toronto
CO			
8 hours (ppm)	28 (92)	24 (91)	13 (91)
Pb			
24 hours (mg/m^3)	0.16 (92)		13.7 (91)
3 months (mg/m^3)	0.09 (92)	2.1 (92)	
NO_2			
1 hour (ppm)	0.28 (92)	0.32 (86)	
annual arithmetic mean (ppm)	0.062 (92)	0.130 (87)	0.066 (91)
SO_2			
annual arithmetic mean (ppm)	0.007 (92)	0.076 (92)	0.01 (91)
O_3			
1 hour (ppm)	0.30 (92)	0.48 (92)	0.163 (91)
PM_{10}			
annual arithmetic mean (mg/m^3)	79 (92)	427 (89)	120 (91)
TSP			
24 hours (mg/m^3)	563 (92)	1494 (89)	379 (91)
annual geometric mean (mg/m^3)	91 (92)	554 (89)	67 (91)

concentration
(year)

Water and Waste at the U.S./Mexico Border: Post-NAFTA Issues*

Randall Crane[+]

Abstract

The US- Mexican border presents some of today's most pressing environmental problems. Rapid urban growth and industrialization along the border has led to large numbers of people are living in areas with inadequate drinking water or solid waste facilities. The continued relaxation of trade barriers between the two countries can only accelerate this process, particularly at transshipment and production centers along the northern Mexican border. This paper summarizes water supply and wastewater treatment issues in the border region, including a discussion of immediate and longer term problems. These issues are then linked to the strategies employed by the infrastructure and environmental policies and institutions now in place.

Introduction

From its inception NAFTA has signified not only the loosening of trade barriers among major trading partners but also the opportunity for Mexico, a developing country, to generally accelerate the process of modernization. This process includes basic elements that go well beyond the simple mechanics of a free market economy, however, such as the effective practice of democracy and the responsibility of government to protect its people and natural resources from the excesses of individual actions. Much of the opposition to NAFTA, both within and without Mexico, centered on these "non-economic" issues. It was not so much the promise of better economic conditions that was questioned as their overall cost and the distribution of that cost. In a sense the broader concern was whether in its rush to leave poverty the Mexican government would compromise its longer-term resource wealth and, on a more personal level, even neglect to bring the poor and the natural environment along.

This paper considers a small but important part of this complex debate, namely planning issues surrounding basic public infrastructure in the U.S./Mexico border region. The topic

[+] Department of Urban & Regional Planning, Department of Environmental Analysis & Design, University of California, Irvine, CA 92717-5150

comprises a diverse set of issues related to trans-boundary environmental problems, local public services for the poor, the role of infrastructure in market-led growth, and the spatial impacts of national economic reform, among many others. It even refers to the role of local governments in the national process of democratization, to the extent political reforms have changed the ability of these jurisdictions to respond to both new constituent complaints as well as more traditional service provision responsibilities (Joint Academies Committee, 1995; Crane, 1995).
The list of border infrastructure issues is not only long, it is often dramatic. The US-Mexican border presents some of today's most pressing environmental problems (e.g., Sepúlveda and Utton, 1984; Fernandez, 1989; Kelly, Kamp, Gregory and Rich, 1991; World Environment Center, 1992). Rapid population growth along both sides of the border have created situations where large numbers of people have inadequate drinking water or municipal solid waste facilities (Pryde, 1986; Alvarez López, 1988; Eaton and Hurlbut, 1992; Orisis, 1992; Sánchez, 1993). Public services on the Mexican side are notoriously poor in many areas — Mexican officials admit that there are only 660 water treatment plants for the country's 92 million people, many not functioning — but serious deficiencies are also in evidence in some rural U.S. communities (Brown and Ingram, 1987).

Most of these problems have little directly to do with trade and even less with NAFTA. They are the natural consequence of rapid urban growth in a poor country, in some instances exacerbated by the special characteristics of a common border with a rich country. Much growth takes place in squatter settlement in marginal or poorly suited areas. Houses are built first with precarious materials, which little by little are replaced by more permanent structures. Electricity is typically the first public service to be installed, often in *jerry-rigged* fashion. Later, running water and sewage are put in place. The period between the settlement and the installation of tap water and sewer can last many years. The concern is that more growth, such as brought about by trade deregulation, will worsen these problems further and also that the opportunities to improve the situation might be foregone.

Nonetheless, NAFTA and related agreements bear directly on these conditions and the strategies for their future improvement. Many observers feel the explicit endorsement of the principle of "sustainable development" in the treaty is a significant and potentially binding gesture. NAFTA otherwise acknowledges a variety of environmental concerns associated with border development and the reduction of trade barriers. In particular, key NAFTA provisions address concerns by environmental organizations to protect and strengthen the enforcement of existing state and federal environmental health and safety standards. As side agreements, the three countries also adopted additional executive agreements dealing with environmental concerns as part of the NAFTA process.

This summary paper surveys these topics in quick succession, with a focus on water and sanitation planning issues in the trans-border region. It begins with a summary of area conditions, with most attention on the widely publicized issues of contaminated river beds, flood control, the manner in which poverty and water access are linked, and appropriate technologies. The paper then looks at the 1991 and 1992 EPA border plans and the NAFTA side agreements addressing these concerns, mainly to describe the entity created to finance infrastructure projects in both sides of the border, the NADBank. Among the top priorities that the bank seeks to finance are sewage and sanitation projects. Current and proposed border water and wastewater treatment projects are summarized. A fundamental element of these planning strategies are their cost effectiveness. Finally, the paper concludes with a discussion of what these developments represent for the basic infrastructure planning process in the border region.

Border Conditions

In most official documents, such as the 1983 U.S./Mexico Border Environmental Agreement, the border area is defined as that within 65 miles on either side of the 2,000 mile common border. This area includes six Mexican states, four U.S. states, and a good number of smaller political jurisdictions. Most of the region is rural and sparsely populated. However, perhaps 12 million people live in 14 pairs of cities. About 75 percent of the U.S. border population and probably a higher proportion of the Mexican border population live in those 14 city pairs. From an environmental perspective, however, the area is undivided in several respects. Many major river basins and desert regions cross or run along the border, and as do groundwater aquifers. The Rio Grande and Colorado River beds run along roughly half the border, for example.
Most border environmental concerns are related to urban and industrial development. Congestion, uncontrolled urban development, and lack of basic public health and sanitation facilities have become significant problems in many communities on both sides of the border. Projected needs for sewage collection and wastewater treatment are perhaps the most acute of the region's environmental problems. Other environmental issues include health risks to residents of neighborhoods where manufacturing firms locate and where, due to rapid urban immigration, people live in close quarters with inadequate water and sewage (Nazario, 1989; Langewiesche, 1992). On the U.S. side, for example, unincorporated communities called colonias have sprung up adjacent to towns and cities. These colonias, which house over 200,000 people in Texas and New Mexico alone, are characterized by substandard housing, inadequate roads and drainage, and barely adequate water and sewer systems, if such systems exist at all (Brown and Ingram, 1989).

In Mexico, thousands of families attracted to the border area by job opportunities in the maquiladora assembly plants have long strained the ability of governments to provide adequate roads, drinking water, and wastewater treatment systems. In some places with high population densities, centralized wastewater collection and treatment systems have never been built. Consequently, untreated or inadequately treated wastewater has been discharged from communities in the border area into rivers, canals, arroyos, the Gulf of Mexico, and the Pacific Ocean. These discharges have contributed to ecological and esthetic degradation, economic losses, and threats to human health. In the Nogales area surface water and shallow drinking water wells have contaminated with pathogenic microorganisms. In the Nuevo Laredo/Laredo area, 27 million gallons per day of untreated wastewater are being discharged directly into the Rio Grande, while treated wastewater fails to meet environmental standards (World Bank, 1994b).

In addition, the hazardous wastes generated by the maquiladoras have caused widespread concern on both sides of the border where little is known about the kinds, quantities, or disposal of such wastes. There are often several direct discharges of raw industrial wastes to rivers and sewers, for example. Difficult terrain in some places has contributed to the excessively high level of pollutants in surface and ground water sources. Yet very little is understood concerning the impacts of the maquiladora industries on border wastewater problems.

The lack of land available for housing, together with the growth of unplanned and largely unregulated informal settlements have generated many communities that are prone to weather related problems. Even small rainfalls in Tijuana can produce severe flooding problems and the major events cause disasters. The flood plain lacks adequate drainage, with water pooling throughout, particularly in the upper area of the Tia Juana River Zone. During the winters of 1980/81, 1983/84, 1991/92 and 1993/94 significant floods occurred, with the loss of at least 37 lives in the last (Dedina, 1995).

One of the most serious environmental problems in the border area is simply insufficient water supply. Rainfall distribution in Mexico is extremely uneven, with severe water shortages in the border area. Average rainfall varies from 100 mm/year in the northern desert regions to 5,000 mm/year in the tropical southern zones. Water supply is severely limited in many of the major border cities. Tijuana, Tecate and Mexicali draw water supplies from the Colorado River; San Luis Rio Colorado, Nogales, Naco, Agua Prieta and Ciudad Juárez rely on grown water; and Ciudad Acuña, Piedras Negras, Nuevo Laredo, Reynosa and Matamoros draw water from the Rio Bravo (see map 1). Ciudad Juárez has perhaps the most acute water scarcity problems. The city draws a major part of its water supply from the Bolson Jueco aquifer, which has a recharge of only 5% of the total annual extraction. Of the 36 aquifers throughout the country that are being overexploited, 21 are located in border states, particularly in the cities of Mexicali, San Luis Rio Colorado and Ciudad Juárez (World Bank, 1994b).

According to the 1990 national census, some 90% of the border population in Mexico has access to piped water, a figure somewhat higher than the national average of 84%. These figures are based on the official figures for the permanent population. Because the number of temporary residents increases the population in the border cities by perhaps 30%, actual water coverage is likely to be substantially lower. Moreover, these figures do not measure the quality of service and include all persons who have water access somewhere in the neighborhood, not necessarily in their home. For example, water supply is often sporadic and the water quality may not comply with national standards. In Ciudad Juárez and Nogales, for example, water is available only one-third of the time (Ingram, Laney and Gillilan, 1995)

While NAFTA is new, relatively open trade in the border region is not. The maquiladora program, an export processing zone for foreign-owned assembly operations, has been in operation for over twenty-five years. This Mexican government program allows foreign firms, most of which have been U.S. owned, to manufacture goods in Mexico with Mexican labor and with foreign materials imported nearly duty free. The U.S. government part of the arrangement allows these firms to export the completed goods back to the U.S. with import tariffs paid only on the value added to the product in Mexico. The program has benefited U.S. firms directly by providing an alternative to distant and more costly Asian labor markets. On the Mexican side, maquiladoras have generated a level of foreign exchange second only to petroleum. From a handful of companies and an employment base of a few hundred workers in 1965, the sector has now grown to over 2,000 firms and nearly 500,000 employees. By any reasonable growth standard, the program has been a success for investors and consumers in the U.S. and many workers in Mexico.

There are other affected groups, however, and growth is not the only relevant standard. Opportunities have been missed on all sides, and growing pains are more pronounced with each passing day. The sector has had only minor stimulating impact on the industrial and technological development of the larger Mexican economy, for example, even in those regions where the operations are concentrated. In this respect especially, the maquiladora assembly operations have disappointed observers and participants alike by remaining an economic and geographic enclave within Mexico. Export processing zones in other developing countries, particularly in the recently industrialized nations of Asia, have served as a springboard for industrial and technological advances, through the strategic development of forward and backward linkages with other home industries. In contrast, the maquiladora industry has not moved much beyond assembly. The learning curve for export-oriented Mexican industry has been less steep than hoped for even while the size of the sector, and the dependence of the Mexican export economy on the sector, continues to grow. The labor force has experienced only moderate transformation and the technological capacity of the nation has not increased in step with employment growth.

In addition, the Mexican economy has in many ways paid dearly for the rapid development of the maquiladora industry. The associated rapid urbanization described briefly above has strained public and private infrastructure beyond capacity and, in some areas, led to declining living and health conditions of tragic proportions. Enforcement of environmental standards is uneven at best, which has had an impact on the health of workers and of the residents living near industrial centers. Some observers have argued that if the maquiladora industry were to close up shop tomorrow, perhaps due to rapid wage growth in the border labor markets, the nation would have little to show for the experience other than border communities that have grown too fast and haphazardly to accommodate their populations.

These concerns are real and legitimate, and they bear tidings for NAFTA, even while the benefits of such efforts have more to offer Mexico in the short run than any other participant. Given the many opportunities presented by the opening of the Mexican economy, it is clear that the benefits promised by free trade should amount to more than semi-skilled job growth and poor residential living standards. Compounding the process of the integration of the Mexican economy with its NAFTA partners are two additional factors: the gap in wealth and wages between the U.S. and Mexico is larger than any two adjacent nations in the world, and the Mexican economy suffered its worst downturns of the century in the mid-1980s and again in early 1995. A fear of dominance and exploitation by the U.S. overlays this contrast. Based on experience, this perception of the U.S. as a threat to both sovereignty and economic integrity has long influenced Mexican attitudes toward a more integrated U.S.-Mexican economic structure.

Questions have been raised regarding the impacts of differences in environmental laws and regulatory enforcement practices have encouraged foreign firms to treat Mexico as a potential site for ‘dirty’ industries. Regarding firm behavior, these concerns take two forms: those related to environmental standards governing industrial production and those related to occupational health standards. Environmental groups on both sides of the border have long complained about the lack of attention and resources provided by either the Mexican government or by maquiladora operators to deal with the pollution and other environmental impacts of industrial development in the border region (Kelly, et al. 1991; Sánchez, 1993; Bath, 1993). It has been reported that some foreign industries establish themselves in Mexico because environmental controls there are less strict, or are less well enforced, than in the U.S. and elsewhere. The U.S. EPA presently requires that all industrial water pollution be treated in municipal systems and in most cases pretreated by large industrial users before being discharged into the municipal system. On the Mexican side, the extent to which the maquiladora operations are responsible for industrial water treatment is subject to an aggressive effort by the government to develop standards on an industry-by-industry basis. Still, virtually no data exist that quantify the scope of the problem.

The prospect of a NAFTA has stimulated new interest in border area conditions, and there is hope that recent official initiatives addressing these issues reflect more genuine enthusiasm than in the past, but environmental programs of one sort or another have been in place for years.

Border Environmental Planning Agreements

The new environmental institutions that came about because of NAFTA were developed in response to the strengths and weaknesses of existing bilateral environmental agreements and treaties. So it is worthwhile to review existing bilateral programs to see how NAFTA changes the nature of the cooperation among the U.S. and Mexico. U.S.-Mexican cooperation on water resources planning and management is based on a diverse set of joint cooperative treaties and activities developed over nearly a century. In 1944 the U.S. and Mexico agreed by treaty to create a new International Boundary and Water Commission (IBWC), extend its authority to the land boundary, and give it lead responsibility for border water sanitation projects mutually agreed to by both countries. In addition to its other duties, the IBWC is currently involved in the planning, construction, operation, and maintenance of several wastewater treatment plans in the region. In 1983 joint Mexican-U.S. environmental activities in the border area were formalized with the signing of a comprehensive Border Environmental Agreement. It established a general framework in which both countries agreed to prevent, reduce, and eliminate sources of air, water, and land pollution. In particular, the 1983 agreement outlined procedures for establishing technical annexes under which specific projects are carried out.

In 1992 the U.S. and Mexican governments announced the creation of 'clean up' funds of $600 million to address environmental conditions in the border region. Though only a beginning, the amount is clearly inadequate to the task, particularly given the expectation of continuing rapid industrial development and population growth in these areas. Another sign of activity on the Mexico side is the reshuffling of environmental and development responsibilities within the federal cabinet, resulting in the creation in early 1992 of a new environmental and development ministry, Secretaría de Desarrollo Social (SEDESOL), replacing SEDUE and consolidating some functions of other ministries. The exact role of the new ministry is not yet clear, but it may signal the kind of commitment needed from the federal government for further progress (Joint Academies Committee, 1995a,b).

The major concern, however, is the commitment of Mexican regulatory authorities to the laws already on the books (Joint Academies Committee, 1995a,b). A significant obstacle to progress is the insufficient funding for enforcement efforts combined with the character of the administrative framework, which reflects the heavily centralized nature of many Mexican institutions. Sufficient discretionary enforcement authority is often not extended to the ranks of the officials in the field, with the result that regulations are applied unevenly and with hesitation. The problem, therefore, is not simply one of funding or the political will of the nation's leaders, but administrative reform in the lower ranks as well. The oft-stated promise of these trade negotiations is that this domestic strength, as well as overall regional growth, will be supported and financed by the gains from more open trade. During the years prior to NAFTA, two processes determined the scope and pace of environmental cooperation along the U.S.-Mexican border: The joint construction activity of the IBWC/CILA and the joint planning efforts of the U.S. EPA and SEDESOL. Since 1993 EPA and SEDESOL have developed drafts of a joint Integrated Border Environmental Plan that has guided efforts by the two federal governments to improve water supply and wastewater treatment among other environmental conditions. But because both processes have relied on federal government initiatives, observers from state and local governments, nonprofit organizations, and the private sector often question the adequacy, efficiency, effectiveness, integrity, or transparency of decisions on border environmental infrastructure (Gilbreath Rich, 1992; Killgore and Eaton, 1995).

The 1983 La Paz agreement lead to the creation of six work groups taking on joint projects related to their names: the Air Work Group, the Water Work Group, the Hazardous Waste Work Group, the Cooperative Enforcement Work Group, the Pollution Prevention Work Group and the Contingency Planning and Emergency Response Work Group. These groups work from Annual Work Plans which focus their energies and direction. IBEP, which covered 1992 to 1994, was an attempt at more long range planning. Border XXI will be the second phase of that long range view, with more emphasis on participation by and input from border communities.

For the 1994-1996 fiscal years the U.S. government has appropriated $778 million through 11 programs in six agencies to fund border environmental institutions and the U.S.-Mexican Border Environmental Plan (Killgore and Eaton, 1995). Both the U.S. and Mexico have pledged to make new unilateral grants for their own communities to upgrade border environmental infrastructure, though it remains unclear just what that entails.

The World Bank had approved $918 million in loans to Mexico for environmental facilities by early 1995, including $368 million to upgrade environmental infrastructure and increase community skills along the Mexican side of the border. Over the period 1994 to 2002, the World Bank's Northern Border Environmental Program loan was intended to provide $225 million for environmental infrastructure though these plans have recently been put on hold. The World Bank may also loan Mexico an additional $1 billion for five other infrastructure projects in Northern Mexico. The IBWC currently is using combined Mexican and U.S. resources to improve treatment capabilities in three sets of sister cities: Nuevo Laredo/Laredo, Nogales/Nogales, and Tijuana/San Diego (see map 2). The IBWC is also working with SEDESOL, EPA and state officials to train IBWC personnel in the proper operation and maintenance of wastewater treatment facilities in the border area.

Concluding Comments

These problems are, for the most part, massive in scale. They reflect the growing pains faced by an industrializing nation. As such, they are not entirely avoidable. The countries on either side of the border have considerable flexibility in the manner in which they choose to manage and accommodate these changes, however. Environmental and infrastructure planning strategies require a combination of resources and political will. By their nature, the private sector will not solve these problems without financial assistance, economic incentives, and strong regulatory guidance by the governments having jurisdiction.

Moreover the public sector likely needs additional administrative assistance together with the right incentives.

Even where governments, businesses and residents agree on a plan of action, many implementation and planning problems remain. The unpopularity of substantially higher prices for water and wastewater treatment is a tremendously difficult institutional and administrative issue, for example. People resist the higher costs for themselves, and in many instances they resist on behalf of the "poor". What is less commonly observed is that many of these problems are linked by the fundamental role of cost-recovery in modern water sector investment practices (World Bank, 1994a; Joint Academies Committee, 1995). Even in those instances where government rhetoric places major importance on extending adequate and affordable services to the poor, services are less likely to be extended to, or improved in, areas which cannot finance them. This view is compounded by the too typical view that infrastructure investment in low-income areas must be financed by subsidies, as self-financing is rejected out of hand as an impractical option (World Bank, 1994a).
In sum, the US- Mexican border presents some of today's most pressing environmental problems. Rapid urban growth has created situations where large numbers of people are living in areas that have inadequate drinking water or solid waste facilities. The continued relaxation of trade barriers between the two countries can only accelerate this process, particularly at transshipment and production centers at the northern border. The extent to which such problems are addressed by the environmental policies and institutions in each area depend not only on a clear understanding of the determinants of those problems, but also on the planning and financial context for those policies and institutions.

Bibliography

1. Alvarez López, Juan (1988) El Medio Ambiente en el Desarrollo Económico de la Frontera Norte de México. Tijuana, B.C.: Universidad Autonoma de Baja California.

2. Argüello, Javier and Michael Najjar (1995) "On the possibility of water trading in the Tijuana/San Diego region," School of Urban and Regional Planning, University of Southern California, November.

3. Bath, C. Richard (1993) "Environment: U.S. perspective," in Sidney Weintraub, ed. U.S.-Mexican Industrial Integration: The Road to Free Trade, Boulder: Westview Press, pp. 318-335.

4. Border Trade Alliance (1992) Southwest Border Infrastructure Initiative: Federal Report, McAllen, Texas.

5. Bradley, David, Carolyn Stephens, Trudy Harpham and Sandy Cairncross (1992) A Review of Environmental Health Impacts in Developing Country Cities, Urban Management Program Discussion Paper No. 6, The World Bank.

6. Briscoe, John (1993) "When the cup is half full: Improving water and sanitation services in the developing world," Environment 35, 7-37.

7. Brown, F. Lee and Helen M. Ingram (1987) Water and Poverty in the Southwest. Tucson: University of Arizona Press.

8. Cairncross, Sandy (1990) "Water supply and the urban poor," in Sandy Cairncross, Jorge Hardoy and David Satterthwaite, eds., The Poor Die Young: Housing and Health in Third World Cities. London: Earthscan Publishers, 109-126.

9. Central Intelligence Agency (1992) "Mexico's northern border: Environmental dimensions," Directorate of Intelligence, Intelligence Research Paper No. IA 92-10015 (August).

10. Clement, Norris and Eduardo Zepeda Miramontes, eds. (1993) San Diego-Tijuana in Transition: A Regional Analysis, San Diego: Institute for Regional Studies of the Californias, San Diego State University.

11. Comision Nacional del Agua (1992) Situacion Actual del Subsector Agua Potable, Alcantarillado y Saneamiento, Mexico City: CNA.

12. Crane, Randall (1993) "Trade liberalization and the lessons of the Mexican maquiladoras," in R. Green, ed., The Enterprise Americas Initiative: Issues and Prospects for a Free Trade Agreement in the Western Hemisphere. Praeger, 83-97.

13. Crane, Randall (1994) "Water markets, market reform, and the urban poor: Results from Jakarta, Indonesia," World Development 22, 1994, 71-83.

14. Crane, Randall (1995) "City planning as nation building: Examples from Mexico," Working Paper, University of California at Irvine.

15. Crane, Randall and Amrita Daniere (1996) "Measuring access to basic services in global cities: Descriptive and behavioral approaches," Journal of the American Planning Association 63 (Spring), 203-221.

16. Dedina, Serge (1995) "The political ecology of transboundary development: Land use, flood control and politics in the Tijuana River Valley," Journal of Borderlands Studies X, 89-110.

17. Eaton, David J. and David Hurlbut (1992) Challenges in the Binational Management of Water Resources in the Rio Grande/Río Bravo. Austin, Texas: The U.S.-Mexican Policies Studies Program, The LBJ School, The University of Texas.

18. Fernandez, Raul A. (1989) The Mexican-American Border Region. Notre Dame, Indiana: University of Notre Dame Press.

19. Fox, Annette Baker (1995) "Environment and trade: The NAFTA case," Political Science Quarterly 110, 49-68.

20. Gray, N.F. (1989) Biology of Wastewater Treatment. New York: Oxford University Press.

21. Gunnerson, Charles (1991) "Costs of water supply and wastewater disposal: Forging the missing link," in H. Rosen and A. Durkin-Keating, eds. Water and the City: The Next Century. Public Works Historical Society, American Public Works Association, Chicago, IL.

22. Ingram, Helen, Nancy K. Laney, and David M. Gillilan (1995) Divided Waters: Bridging the U.S.-Mexico Border. Tucson, AZ : University of Arizona Press.

23. Kabelmatten, J.M., D.A. Julius and C.G. Gunnerson (1982). Appropriate Sanitation Alternatives- Technical and Economic Appraisal. The World Bank Washington, D.C.

24. Kelly, Mary, Dick Kamp, Michael Gregory and Jan Rich (1991) "U.S.-Mexico free trade negotiations and the environment: Exploring the issues," The Columbia Journal of World Business 26, 42-58.

25. Killgore, Mark and David J. Eaton (1995) NAFTA Handbook for Water Resource Managers and Engineers. The U.S.-Mexican Policies Studies Program, The LBJ School, The University of Texas, and The American Society of Civil Engineers

26. Joint Academies Committee on the Mexico City Water Supply (1995a) Mexico City's Water Supply: Improving the Outlook for Sustainability, A National Research Council Report. Washington, D.C.: National Academy Press.

27. Joint Academies Committee on the Mexico City Water Supply (1995b) El Agua y la Ciudad de México: Abastecimiento y Drenaje, Calidad, Salud Pública, Uso Eficiente, Marco Jurídico e Institucional. Mexico City: Academia de la Investigación Científica, Academia Nacional de Ingeniería, and Academia Nacional de Medicina, in collaboration with the National Academy of Sciences.

28. Langewiesche, William (1992) "The Border," The Atlantic Monthly (June), 91-108

29. Middlebrooks, E.J. and S. Reed (1995) "Water stabilization ponds on the Mexico-USA border. Performance and upgrading potential," Draft report presented to the US EPA, September 1995.

30. Nazario, Sonia (1989) "Boom and despair: Mexican border towns are a magnet for foreign factories, workers and abysmal living conditions," Wall Street Journal (September 22), p.1.

31. Orisis, D. (1992) A Study of the Health Impacts of Storing Potable Water in Drums: The Case of Tijuana, Mexico, unpublished M.A. thesis, SDSU.

32. Postel, Sandra (1992) Last Oasis: Facing Water Scarcity. New York: W.W. Norton & Co.

33. Pryde, Philip (1986) "A geography of water supply and management in San Diego-Tijuana," in L.A. Herzog, ed. Planning the International Border Metropolis: Trans-Boundary Policy Options for the San Diego-Tijuana Region. San Diego: Center for U.S.-Mexican Studies, University of California at San Diego.

34. Rich, Jan Gilbreath (1992) "Planning the border's future: The Mexican-U.S. integrated border environmental plan," Occasional Paper No. 1, The U.S.-Mexican Policies Studies Program, The LBJ School, The University of Texas.

35. Sanchez (1995) The San Francisco Chronicle. Friday, September 29, 1995 page D4 "Nafta Panels on Environment Decide in 2 cases."

36. Sánchez, Roberto A. (1990) "Health and environmental risks of the maquiladora in Mexicali," Natural Resources Journal 30, (Winter) pp. 163-186.

37. Sánchez, Roberto A. (1993) "Environment: Mexican perspective," in Sidney Weintraub, ed. U.S.-Mexican Industrial Integration: The Road to Free Trade, Boulder: Westview Press, pp. 303-317.

38. Sánchez, Roberto A. (1993) "Teoría y método para el estudio de los servicios urbanos en la frontera norte de México: El caso del agua," in Urbanizacion y Servicios. El Colegio de la Frontera Norte y Universidad Autónoma de Ciudad Juárez, 7-19.

39. Sepúlveda, César and Albert E. Utton, eds. (1984) The U.S.-Mexico Border Region: Anticipating Resource Needs and Issues to the Year 2000, El Paso: Texas Western Press, The University of Texas at El Paso.

40. Sierra Club (1993) Funding Environmental Needs Associated with the North American Free Trade Agreement, Washington, D.C. (January).

41. Texas Center for Policy Studies (1992) NAFTA and the U.S./Mexico Border Environment: Options for Congressional Action, Austin, Texas (February).

42. Texas Center for Policy Studies (1994) Fulfilling Promises: Implementation of the Border Environment Cooperation Commission (BECC) and the North American Development Bank (NADBANK), Austin, Texas (February).

43. Tijuana, Ayuntamiento del Municipio de (1994) Programa de Desarrollo Urbano del Centro de Población Tijuana, Tijuana, B.C.: Direccion de Planeacion del Desarrollo Urbano y Ecologia (Julio).

44. Trava Manzanila, José, Jesús Román Calleros and Franciso A. Bernal R., eds. (1991) Manejo Ambientalmente Adecuado de Agua en la Frontera México-Estados Unidos. Tijuana, B.C.: El Colegio de la Frontera Norte.

45. Council of the Mexico-U.S. Business Committee (1993) Analysis of Environmental Infrastructure Requirements and Financing Gaps on the U.S./Mexico Border, Washington, D.C. (August).

46. and S.E.D.U.E. (1992) Integrated Environmental Plan for the Mexican-U.S. Border Area (First Stage, 1992-1994). Washington, D.C. and Mexico City.

47. (1994) Water Pollution. Information on the use of alternative wastewater treatment systems. US. General Accounting Office. GAO/RCED-94-109. EPA (1995). Compendium of EPA Binational and Domestic US/Mexico Activities. EPA 160-B-95-001

48. Trade Representative Office (1994) Environmental Impact of NAFTA. Rockville, MD: Government Institutes.

49. Whittington, Dale and Venkateswarlu Swarna (1994) The Economic Benefits of Potable Water Supply Projects to Households in Developing Countries, Economic Staff Paper No. 53, Manila: Asian Development Bank, January.

50. World Bank (1980) Poverty and Basic Needs Series: Water Supply and Waste Disposal, Washington, D.C.

51. World Bank (1993) World Development Report, 1993: Investing in Health. Washington, D.C.: Oxford University Press.

52. World Bank (1994a) World Development Report, 1994: Infrastructure for Development. Washington, D.C.: Oxford University Press.

53. World Bank (1994b) Mexico: Northern Border Environmental Project, Staff Appraisal Report No. 12603-ME, Washington, D.C. (May).

54. World Bank (1994c) Mexico: Second Water Supply and Sanitation Sector Project, Staff Appraisal Report No. 12340-ME, Washington, D.C. (May).

55. World Environment Center (1992) Environmental, Health, and Housing Needs and Nonprofit Groups in the U.S.-Mexico Border Area, Arlington, Virginia (June).

* Abridged version of an invited paper for the NSF conference "Environmental Quality, Innovative Technologies, and Sustainable Economic Development — A NAFTA Perspective" held at UNAM in Mexico City, February 8-10, 1996. I am grateful for the comments of Alberto Pombo and Charles Gunnerson on earlier versions.

These cities are: Matamoros, Tamaulipas, and Brownsville, Texas; Reynosa, Tamaulipas, and McAllen, Texas; Nuevo Laredo, Tamaulipas, and Laredo, Texas; Piedras Negras, Coahuila, and Eagle Pass, Texas; Ciudad Acuña, Coahuila, and Del Rio, Texas; Ojinaga, Chihuahua, and Presidio, Texas; Ciudad Juarez, Chihuahua, and El Paso, Texas; Las Palomas, Chihuahua, and Columbus, New Mexico; Agua Prieta, Sonora, and Douglas, Arizona; Naco, Sonora, and Naco, Arizona; Nogales, Sonora, and Nogales, Arizona; San Luis Rio Colorado, Sonora, and Yuma, Arizona; Mexicali, Baja California, and Calexico, California; and Tijuana, Baja California, and San Diego, California.

Sustainable Development and Environmental Conservation in the Americas

Raúl A. Deju*

The 1990's have been a period in the history of our continent that started the era of regional environmental cooperation. Certainly, NAFTA, created a cooperative structure including the North American Commission for Environmental Cooperation (NACEC) headquartered in Montreal, Canada as well as numerous institutional linkages at various levels.

Subsequent to NAFTA, the process toward free trade throughout the Continent by 2005 as exemplified in the declaration of nations at the 1994 Miami Summit of the Americas and the 1995 Denver Hemispheric Trade and Ministerial Forum, commits the nations of the continent to:

- Promoting Prosperity through economic integration and free trade and
- Guaranteeing sustainable development and conservation of the environment.

These two principles are not necessarily opposites. Economic prosperity that irreparably harms the environment is ultimately not prosperity but a step backward.

Sustainable development and environmental conservation need not be a drain on the national treasuries nor to the businesses and ultimately the people in our continent. A gradual environmental improvement program tied to the growth in domestic production and supplemented by training and enforcement actions is to me a preferred option over a quick and expensive program that has no public support nor public understanding.

Business leaders, public interest groups, academics and government officials within each nation need to be involved in developing the environmental vision for each of their nations.

* President DGL International, Inc., P.O. Box 3491, Walnut Creek, CA 94598 & Professor of Environmental & Resource Sciences, University of California at Davis

From my perspective as a business executive and an academic the following ten steps should constitute a comprehensive Sustainable Development and Environmental Conservation Program that also ensures Economic Integration:

(1) Each country in the Americas should have in place a strong environmental training and a strong enforcement program. Enforcement of environmental laws is each country's responsibility and right under national sovereignty principles. Enforcement programs will likely be different in each country, however, these differences should not create a situation where environmental laws and their enforcement in a given country create adverse transboundary impacts to other countries in the Americas or provide a given country and its industry unfair trade advantages. Training and enforcement programs should draw as much as possible from regional cooperation.

(2) Enforcement of environmental laws in the Americas should include sufficient coordination and cooperation between nations and encourage joint enforcement programs between border nations; a clear continental treatment for problems that cross national boundaries; information sharing initiatives; and mechanisms for rapid deployment of technical assistance, technology transfer and training between the nations of the Americas.

(3) A program should be put in place to (a) promote the gathering and dissemination of information regarding overall national levels of emissions and major pollutant discharges to the ambient environment; (b) promote an environment of consistent compliance monitoring, and (c) report (in comparable terms) on the progress toward achieving sustainable development and environmental conservation including information on compliance investment levels in each country, compliance rates and the results of enforcement actions.

(4) Each country of the Americas consistent with Agenda 21 and the Framework Convention on Climate Change should promote sustainable energy development and use. Such a program should include development of a least-cost national energy strategy. Each country should encourage the private sector and multilateral financial institutions' involvement in the development of their energy resources.

(5) Each country should seek to ratify and begin implementation of the provisions of the Framework Convention on Climate Change which entered into force on March 21, 1994.

(6) Each country of the Americas consistent with Agenda 21, the Convention on Biological Diversity, and other related international instruments should ensure that strategies for the conservation and sustainable use of biodiversity are integrated into relevant economic development activities such as forestry, agriculture, manufacturing, tourism, coastal zone management and fishing.

(7) The nations of the Americas should build an environmental protection framework that utilizes existing institutions minimizing the need for creation of additional bureaucracies that do little for environmental protection yet increase the operating environmental management budget for each nation unnecessarily.

(8) The focal point for environmental protection for each country in the Americas particularly the developing ones should be to implement water and sanitation projects to eradicate within the next 20 years diarrheal disease, typhoid, schistosomiasis, amebiasis, hookworm, hepatitis A, ascariasis, giardiasis, trichuriasis, dracumuliasis and problems related to ingestion of water contaminated with organic or inorganic compounds. Water treatment projects should be implemented in a manner that ensures safe drinking water to all citizens of the continent.

(9) The countries of the Americas should standardize environmental protection and environmental infrastructure requirements such that any infrastructure required generates either an in-country business opportunity or a trade opportunity for a sister nation in our continent.

(10) The countries of the Americas should make use of private sector expertise, academic institutions and talent from other nations in the continent to speed up solutions of environmental issues while keeping costs down.

The countries of the Americas should cooperate to define and quantify the universe of sustainable development and environmental conservation challenges, as best they can, before designing extremely expensive programs and allocating scarce resources. The ten points raised above should constitute the core of any national program.

Mexico's Needs

Where does Mexico fit in this context? is a question I would like to discuss at this time. As one of three NAFTA signatories, Mexico has committed to extensive regional environmental cooperation. Mexico has also committed to the goals of Agenda 21 and in the past 5 years has taken tremendous steps toward enhancing environmental enforcement, fostering cleanups, and improving the environmental conscience of its industries.

Mexico's biggest need to date is in the area of modernizing its environmental infrastructure. The modernization of this infrastructure needs to go hand in hand with the improvements in the Mexican economy.

Mexico needs to both modernize the production machine of the country including an ability to rapidly move goods and services while at the same time improving the quality of life and environment in Mexico by providing clean water, clean air, waste treatment, adequate housing, electricity, phone, transportation, sanitation and waste disposal.

The development of this environmental infrastructure is key to achieving the country's economic development goals and viceversa. Infrastructure needs to be defined as not just a bunch of physical assets but also the accompanying institutional strengthening necessary to get the job done.

Environmental infrastructure financing and development will provide an opportunity for multilateral and private financial institutions in Mexico and abroad, construction and environmental companies and owners of applicable technologies both within and outside of Mexico to participate in a large investment market. To finance the infrastructure needed, a great deal of which is needed in the border region between the U.S. and Mexico, and given the existing deficiency in Mexican aggregate savings implies a need to consider a variety of options such as:

- Financing from the World Bank, the Interamerican Development Bank and other multilateral agencies;
- use of user charges and fees coupled with long-term concessions;
- development of public-private partnerships;
- use of financing from Mexican Banks, especially its national development banks; and
- use of government guarantees.

Public institutions and companies from outside Mexico desiring to participate in this major endeavor need to be creative, practical and understanding of the cultural differences between Mexico and their respective countries. Mexico for its part needs to continue to encourage the flow of foreign capital needed for infrastructure projects. To date, Mexico has shown a great deal of progress through:

- extensive liberalization of laws designed to limit private and foreign involvement in parts of the Mexican economy;
- provision of most favored nation status to American and Canadian companies as part of NAFTA;
- development of rules that clearly spell out the process of expropriation and limit it considerably;
- development of rules that spell out a sensible dispute resolution process; and
- development of rules that do not limit currency convertibility and allow repatriation of profits.

Further progress is needed to address mechanisms to minimize investors risk from long-term financing of cross-border projects including risk from currency depreciation, taxation changes, regulatory changes and enforceability of claims.

Two key entities that will help develop the needed environmental infrastructure in the U.S.-Mexico border are the North American Development Bank (NADBank) and the Border Environmental Cooperation Commission (BECC). The NADBank is expected to play a significant role in the financing of environmental infrastructure projects in the Mexican Border. The legislation funding NADBank by the U.S. and

Mexico provides 90% of the NADBank's loans and loan guarantees be targeted to border environmental projects in the U.S. and Mexico. Ten percent of the funds the U.S. contributes could be earmarked for communities negatively affected by NAFTA anywhere in the U.S. NADBank alone, however, is not the answer. In months to come the challenge is to encourage savings in Mexico, develop in Mexico a finance industry and fully stabilize the national currency. When these steps are in place, investors will be able to participate in environmental infrastructure projects in a realistic fashion with proper consideration of the desire of the constructors and the financial backers to receive a reasonable profit and minimize risk. Projects that do not meet profit requirements but are needed to meet social needs should be subject of government financing partially or totally.

A second challenge is to evolve a clear and simple relationship between NADBank and the BECC. The key is to have the priorities for financing well defined and to insure proper local support and environmental viability for each project being funded to make certain they get done without undue roadblocks.
The BECC has been tasked with figuring out how to spend the NADBank money. However, it will also be able to help communities put together projects that would be marketed outside the NADBank, and this is likely to involve many public-private partnerships.

The BECC will be important because no institution like it currently exists to facilitate cross-border environmental planning and development. It will also be important because it will draw the federal government of both countries into the border environmental infrastructure planning process at an early stage and will be in a position to help bundle and otherwise make attractive projects that by themselves would probably not be salable. The initial BECC structuring is encouraging.

I believe that through the effort of both the U.S. and the Mexican governments, through NADBank/BECC, through NAFINSA and BANOBRAS, through the effort of the World Bank, the Interamerican Development Bank, and other multilaterals and through private financing, investors will obtain the needed financial backing for many projects.

Now, let's address the size of the Mexican environmental infrastructure market. In the last 3 years Mexican government estimates are that spending for environmental infrastructure projects was about 1% of Mexico's gross domestic product. Spending for industrial environmental improvements is likely to increase at 10-15% per year.

While it is difficult to exactly forecast what areas are likely to receive attention, it is expected that municipal wastewater projects and air pollution control projects will lead the list. The environmental infrastructure needs for Mexico clearly fit within the continental priorities noted in the preceding pages.

Air pollution control equipment vendors have an excellent growth market in the valley of Mexico and in the major cities around the country. As PEMEX upgrades and reformulates its gasoline and other products it can generate a booming market for construction and technology companies to transform and modernize the Mexican petroleum industry.

Solid waste concessions are being bid and awarded in various municipalities across the country. A number of hazardous waste projects are underway. Various international companies have taken steps to work with Mexican companies to address the building of solid waste/hazardous waste infrastructure projects. Solid and hazardous waste investments are likely to be substantially outpaced by the municipal waste water treatment investments.
Another area where investment is likely is in the modernization of the Mexican manufacturing infrastructure. As plant and equipment is modernized it presents an opportunity to incorporate "green" design considerations into the re-design.

Mexico's environmental needs require both timely and cost-effective solutions. Through regional cooperation Mexico can greatly reduce the time it takes to improve new technologies and pick some winners that are cost-effective. Mexico will not likely be a test bed for unproven technologies. In addition, through regional cooperation Mexico can implement cost-effective education, training, and enforcement initiatives.

Mexico's progress in the environmental arena the past 5 years has been remarkable considering the extended fiscal crisis of the past 2 years. Mexico, in my mind, has shown the will to achieve sustainable development and now needs to lead the rest of the Americas on the path towards further improvements.

CONCLUSION

Sustainable development and environmental conservation are wise social goals for all nations. Specific projects which may be required need to be prioritized and their cost effectiveness ensured. It may take two decades or more to achieve comparable programs throughout most of the continent, however, the important point is to begin moving on a path that will get us there. Mexico has already demonstrated the value of cooperation especially as a result of NAFTA. Let's expand cooperation to encompass the entire continent and make our continent a better place for our future generations. Let's not decimate our resources such that our children and grandchildren will pay the price for our folly. I strongly urge all of us to remember the words of one of this century's greatest dreamers, "Walt Disney". A phrase he commonly used is very applicable to environmental conservation and sustainable development: "If you can dream it, you can do it." Let's get started.

Structuring a Collaboration between Mexico and the U.S. for managing Water Resources of mutual interest in the NAFTA context

Dennis Engi*

Proper management of water resources is clearly essential to economic activity which, in turn, is fundamental to trade among Canada, Mexico, and the United States. In 1944 the United States and Mexico signed the Colorado River Treaty defining water rights along the leading source of fresh water in much of the Southwest. In the United States there now are numerous diversions including 17 major dams that drain the river's water for use by more than 21 million people in seven states. In additions to the obvious quantity of supply issue, water quality issues are also of significant importance to Mexico and the United States for the Colorado River as well as other water supply options. To address the myriad of important water resources management issues a new 'framework' is needed to ensure appropriate representation of the salient institutional perspectives on both sides of the border.

Recently, a decision support system was structured for water resources management of the Middle Rio Grande Basin in New Mexico. The framework used for structuring this system is suggested as a good 'model' to structuring a decision support system for water resources management in the context of NAFTA. The following discussion describes the result of the Middle Rio Grande Basin experience in structuring a decision support system for water resources management.

The Middle Rio Grande Basin Water Resources Initiative is intended to provide extensive, reliable, and useful information to support water resources management decision making in the Middle Rio Grande Basin into the 21st century. This activity involves numerous collaborating participants including federal, state, and local government agencies; New Mexico universities; and private corporations. The goal of the Initiative, as determined by an exceptionally diverse panel of stakeholders assembled by Sandia National Laboratories, is to create a state-of- the-art information collection, analysis, and graphic delivery system useful to decision

* Sandia National Laboratories, Alburquerque, NM

makers that describes, predicts, and depicts the availability and distribution of water in the Middle Rio Grande Basin.

A set of six water resources management issues for the Middle Rio Grande Basin defined by a second stakeholder panel representing a variety of perspectives including government, industry, public interest groups, and academe includes: (1) responses of the aquifer, Rio Grande, and biota to natural processes and human activities; (2) values to guide water resources decision making; (3) water requirements to satisfy demands, including the population, and the economy; (4) lack of consensus on the water resources budget for the Middle Rio Grande Basin; (5) water resources allocation and management; and (6) legal controls, jurisdictional considerations, and management impacts on the river and aquifer.

Specific information needs under each of these vital issues have been identified by a third expert panel of stakeholders. These information needs range from technical, such as hydrogeologic data, to socioeconomic, such as future water use preferences and industrial growth projections, and will guide the information collection, analysis, and dissemination activities of the Initiative. The following illustrates the range of information needs identified by the expert panels:

Response of the aquifer, river, and biota to natural processes and human activities basic aquifer parameters (e.g., storage coefficients, hydraulic conductivity) compilation of existing water resources data, including historical records contaminant transport and distribution geo-hydrological framework and subsurface structure interconnection between the river and aquifer natural and artificial recharge information roles of specific species in the riverine ecosystem subsidence surface water flows vertical aquifer characteristics water quality with depth data water table and hydraulic head data.

Values to guide water resources decision making acceptable water costs conservation practices education programs Indian pueblo interests land use practices locally relevant information public opinion on water issues sustainability of water resources water management goals.

Water requirements to satisfy demands (e.g., population, economy, ecology) current water requirements, withdrawals, and consumptive use industrial growth projections population growth projections projected water needs.

Lack of consensus on water resource budget for the Middle Rio Grande Basin basin modeling of groundwater and surface water flows return flows to the river surface and groundwater supplies water consumption and losses water inflow, outflow, and storage information water quality impacts on water quantity.

Water resource allocation and management acceptable decision making processes basin yield optimization resource management practices.

Legal controls, jurisdictional considerations, and management impacts current water use obligations and restrictions jurisdictional road map legal framework water rights status.

Increased pressure on water resources resulting from continued population and economic growth has made the absence of a widely agreed upon technical basis for decision making evident to virtually all stakeholders. Recent models of groundwater resources, particularly the hydraulic connection between the aquifer and the Rio Grande, conclude that the highly productive zone of the aquifer is much less extensive and is thinner than previously believed. These models represent an uncertain tool for decision making until the nature of the water budget and the dynamics of the aquifer and river system are better understood. A particularly important contribution of the Initiative is the collection of key information to allow more extensive validation of such models.

In the absence of developing agreement on the basic facts characterizing the aquifer and Rio Grande system, the past practice of resolving differences by informed discussion and debate will be replaced by adjudication, an extraordinarily expensive and lengthy legal process. An immediate commitment to the Initiative is needed to prevent such divisive actions from undermining the required cooperative efforts to compile information to support water resources decision making.

Output of the Initiative will be made widely available at a single point of entry, while maintaining the integrity of existing information systems, through creation of an electronic water information network containing all relevant technical and socioeconomic data. Maintenance of the network will be assigned to an objective organization mutually agreed to by all the concerned parties.

The Middle Rio Grande Basin experience clearly demonstrates a viable framework for structuring a decision support system for water resource management in the context of NAFTA. Proper implementation of such an approach ensures the inclusion of the spectrum of stakeholder interests in managing this most valuable resource.

HEALTHY ENVIRONMENTS AND ENVIRONMENTAL HEALTH

Howard Frumkin*

One of the many challenges posed by NAFTA is our language differences. The major languages of the NAFTA countries, of course, are Spanish, French, and English, and there are dozens of others from Inuit in the north to Mayan dialects in the south. But in this presentation I want to address another language difference, perhaps more difficult than the ones just mentioned: the difference between engineering-speak and health-speak.

Unfortunately, the environmental professions such as environmental engineering have developed almost completely independently from the health professions. For the most part we train in different institutions. We belong to different professional organizations. We read different books and journals. We define our paradigms differently.

This is unfortunate because environment and health are inextricably linked. First, the environment affects health, a point that is by now generally accepted. Second, in the eyes of the public and policy-makers, a principal driving force behind environmental concern is human health. Most people prefer thick forests and clean rivers, but many would sacrifice them for economic reasons. However, very few people would sacrifice the health of their children, if that be the cost of environmental contamination.

I believe that environmental engineers, and others responsible for designing and achieving cleaner, greener, technologies, and health professionals, responsible for assessing, treating, and preventing health hazards, ought to work to overcome their linguistic differences. We will do our work better, and we will make our case more successfully to the public and to policy-makers, if we work together. Accordingly, as one of the few health professionals in this distinguished gathering of environmental experts, I want to outline eight basic perspectives of environmental health, and discuss their interface with the other environmental fields represented at this workshop.

1. **Environmental exposures cause a wide range of health effects.** Over the last two decades, enormous attention has been focused on cancer. In fact, the entire enterprise of quantitative risk assessment grew out of a cancer paradigm. But we must not forget the lessons of history--that environmental exposures can cause respiratory disease, kidney disease, neurological disease, and others. We are increasingly focusing on subtle neurological deficits, such as impaired development in children with relatively low levels of lead exposure. We are increasingly appreciating the effects of certain classes of chemicals

* M.D., Dr.P.H., Department of Environmental and Occupational Health, Rollins School of Public Health, Emory University, Atlanta, Georgia, 30322

on the endocrine system, affecting reproduction and other functions. Immune function may be affected by environmental toxins. To these must be added the physical effects that are mostly confined to a specific type of environment, the workplace: musculoskeletal disorders of the arms, the neck, and the back. In designing environmentally safe and healthy technologies, we need to consider the many dimensions of safety and health in terms of human outcomes.

2. **These health effects may be acute or chronic.** A sudden exposure to an irritating gas such as ammonia will cause an immediate reaction, with watery eyes, irritated mucus membranes, and difficulty breathing. Skin exposure to hydrogen fluoride will cause serious tissue damage within hours. Indeed, much of the environmental safety field has traditionally focused on these acute outcomes. A glance at almost any Material Safety Data Sheet will show far more information on acute toxicity than on delayed or long-term effects. But these long-term effects, such as cancer and endocrine disruption, are equally important. In designing environmentally safe and healthy technologies, we need to consider both short and long time frames.

3. **People are heterogeneous in how they respond to environmental hazards.** Here is a central difference between engineering and biomedical thinking. The engineer thinks in terms of predictability; given a specific design and construction, and a specific set of conditions, equipment should function in a predictable and consistent manner. The physician or epidemiologist, in contrast, thinks in terms of variability. If we turn up the heat in this conference room, some of us will be uncomfortable, and others will not. If we prepare our next meal with monosodium glutamate to enhance the flavor, some of us will end up with headaches, others may wheeze, and others will simply call home to rave about the food in Mexico City. How does this heterogeneity manifest? In some cases, the population response to a hazardous exposure is normally distributed; all of us are reactive, most at some intermediate level, some of us with a high level of reactivity, and some of us nearly invulnerable. In other cases, the population response is bifurcated; allergic or otherwise sensitive people will react, while others will not. In still other cases, we know that certain subpopulations have special vulnerability. For example, when air pollutants climb to high levels in Mexico City, Los Angeles, or Sao Paulo, the primary victims come from three groups: the very young, the very old, and those with pre-existing cardiac or respiratory disease. In designing environmentally safe and healthy technologies, we need to design not for a mythical mean, but for the most susceptible groups in our societies.

4. **Hazardous environmental exposures occur in combinations, rarely in isolation.** early all of our experimental data, and all of our theoretical calculations, consider pure exposures to single hazards. This is fine for laboratory exercises, but not very informative for real-world applications. We may make careful projections of the adverse health effects of an increase in ozone from 0.09 to 0.14 ppm, but what about when this occurs together with exposures to oxides of sulfur and nitrogen, particulates, pollen, and organics? In designing environmentally safe and healthy technologies, we need to remember that protection against multiple, often simultaneous hazards, is required.

5. **Environmental health and occupational health are closely linked.** In fact, the workplace is a uniquely informative environment, since many hazardous exposures occur there at higher levels, but otherwise in similar ways, to those in the general environment. This principle, from the health side, has exact correspondence on the engineering side. A

ventilation system in a factory, at a work station, and a scrubber at the end of the smokestack, rely on similar physical principles. Technical advances both in the workplace and in the general environment draw on a shared scientific base, need similar kinds of expertise, and should proceed in parallel. However, this point is not widely appreciated by policymakers, since labor issues and environmental issues are almost always assigned to distinct agencies, supported by distinct constituencies, and governed by distinct regulations. In designing environmentally safe and healthy technologies, we need to think in terms of an entire systems approach, from the point of production to the point of use, anticipating related health effects throughout.

6. **A wide range of methods is used to study health effects of environmental exposures.** Epidemiology studies patterns of disease in human populations. Epidemiologic studies are extremely informative since they focus on the element of direct interest, humans in the real world. However, the real world includes a multitude of confounding factors--smoking, drinking, variable self-reporting, missing medical records, and so on--that deprive epidemiology of the precision and control of laboratory studies. Small clinical studies may experimentally expose volunteers to hazards, providing useful information, but only about acute reactions. Animal studies may also expose subjects, for either short-term or long-term periods. Here far more precision can be achieved, but extrapolation from animals to humans always involves a certain leap of faith. Finally, laboratory studies with cell systems are used, especially in screening for mutagenicity, which is thought to predict carcinogenicity. Again, the cost of increased precision of the laboratory is the need to extrapolate from cells systems to humans, with all the uncertainty that process implies. When designing environmentally safe and healthy technologies, we need to draw on a range of data sources to evaluate potential hazards, understanding the strengths and limitations of each source.

7. **Primary prevention is the preferred approach to environmental health hazards.** This is a principle shared by environmental engineers, who understand the advantages of source reduction over end-of-pipe control, and by public health professionals, who know that an ounce of prevention is worth a pound of cure. In designing environmentally safe and healthy technologies, we need to abide by principles of primary prevention.

8. **Quantitative risk assessment is a useful technique, but one with many important limitations.** Quantitative risk assessment (QRA) has become a standard approach to environmental health regulation. In the U.S., a four-step method was formally defined over a decade ago, including hazard identification, exposure assessment, dose-response assessment, and risk characterization (National Research Council, 1984). Certainly, the urge to quantify health risks, to rank them, and to define how much risk is acceptable, is laudatory. But QRA is subject to important limitations. First, it is based on extrapolation from animals or from high-dose human exposures to low-dose human exposures, an extrapolation that rarely has a quantitative basis. Second, QRA entails an almost exclusive focus on cancer, while other health endpoints such as reproductive toxicity may be more important. Third, QRA is usually based on data for isolated toxins, while real human exposures, as noted above, are typically multiple. Fourth, QRA results appear to be quantitative and precise, but are actually extremely sensitive to assumptions; risk estimates can vary by seven or eight orders of magnitude. Finally, QRA raises disturbing questions of equity and fairness, since those who calculate risk, and determine how much risk is

acceptable, are usually not those who sustain the risk. This is a problem with no easy answer, since there is a real need to assess the magnitude of risks in setting policy. But as we design environmentally safe and healthy technologies, we should beware of the false precision of QRA data, and constantly strive for the lowest risk we can achieve.

What does this all mean in the context of NAFTA? Companies from the U.S. and Canada (and elsewhere) are attracted to Mexico by relatively cheaper labor costs, and by what they perceive as a more permissive regulatory environment. This creates several kinds of pressures. Governments at the Federal, state and provincial, and local levels in the U.S. and Canada face pressure not to regulate too strictly, lest they increase the pressure for firms in their jurisdiction to move south, taking jobs with it. Consequently, environmental health threats in the U.S. and Canada are aggravated. This is now more true than ever, especially in the U.S., with enormous political pressure being applied against OSHA and EPA. Both enforcement agencies have seen funding cuts, staffing cuts, and legislative pressure to restrict their enforcement. Small companies especially, which employ the majority of the workforce, often lack the expertise and resources to implement clean, green production techniques. In the current political and economic climate, there is little incentive for them to invest in clean production, and great competitive pressure to cut corners.

At the same time, companies that move to Mexico may operate less strictly than in their home countries, creating environmental health threats in Mexico. Who is affected? Certainly, Mexican workers are affected. For example, the maquiladora plants, both along Mexico's northern border and increasingly in the interior, employ more than 500,000 workers in labor-intensive jobs, manufacturing electronic equipment, textile products, automobile parts, toys, and other products (Frumkin et al., 1995). Exposure to chemical hazards--solvents, acids and bases, metals, and other toxins-- as well as physical hazards--repetitive motion, inadequate lighting, noise, and heat--may be common. In addition, the general population is affected by discharges from the plants, and by secondary exposures that workers bring home on their clothing. Maquiladoras are not the only facilities that deserve concern. Domestic plants in Mexico, especially smaller companies, are just as under-resourced as their counterparts north of the border. One visible example is the brickmakers, or ladrilleros, who are busy producing bricks to supply the building boom along the border. The ladrilleros are micro-enterprises, often using primitive technology such as home-made ovens that burn old tires and other trash as fuel. Every day, dozens of black plumes high in particulates, oxides of nitrogen, ozone, and organics rise above Ciudad Juarez from the ladrilleros' operation. Such economic activity, and the environmental threats to health it generates, must be seen as directly linked to NAFTA.

A full account of environmental health hazards in the context of NAFTA is beyond the scope of this paper. Moreover, since available data on exposures and their health effects are limited, a truly full account could not now be written. However, we do know that all major media are affected. These include a variety of exposures within the workplace, air pollutants from large and small factories, power generation, and automobiles and trucks, emissions into waterways, and hazardous and solid waste generation. Even pesticides in the rural environment and in the food chain are affected by international economic integration, since the pesticides become more widely available to farmers throughout the trade area, and the food that results, possibly contaminated with pesticides, is also sold throughout the region.

What is to be done? To advance environmental health under NAFTA, several needs can be identified.

First, we need better health data. Health surveillance of working populations, the general population, and special subpopulations such as children, and exposure data in various environments, must be generated consistently and made available. At present, such data are collected incompletely, inconsistently, and by a huge variety of agencies, organizations, and companies, which sometimes maintain them confidential. Without good public health surveillance, we cannot identify and target priorities for intervention. Second, we need consistent practice with regard to environmental health throughout the region. One of the great disappointments of NAFTA is that it is only a trade agreement. Unlike the analogous process in Europe, it does not harmonize environmental and occupational standards upward among member countries. Even if the three NAFTA nations do not do this as a matter policy, transnational companies should do so as a matter of business ethics. There is no excuse for a company operating in a dirtier, less protective manner in Mexico than in the U.S.

Third, and directly implied by the first two points, we need more human resources. We need to train epidemiologists and environmental health specialists, just as we need to train environmental engineers and industrial hygienists, across the NAFTA region. Especially here in Mexico, such professionals are in critically short supply. Professional organizations and academic institutions have a major role to play in internationalizing environmental health expertise, keeping pace with the internationalizing of business itself.

Fourth, we need a free and open flow of information. In the U.S. we recognize a Right to Know about hazardous exposures, for both workers in the workplace and members of the public. This right should be widely recognized, expanded, and put into action. It is essential to creating an informed and active public, which in turn is essential to creating good environmental health practice. In fact, NAFTA creates unique opportunities for uniform information flow in all the member countries.

Fifth, we need aggressive technological innovation. NAFTA provides unique opportunities for such innovation. To cite an earlier example, the ladrilleros of Ciudad Juarez, a fascinating collaboration of a Mexican NGO, FEMAP, business owners on both sides of the border, Sandia National Laboratory, and the El Paso Gas Company, along with several government agencies, has led to the development of more efficient ovens, using natural gas instead of trash as fuel, which are beginning to replace the more primitive polluting facilities. This could only have been accomplished in the context of NAFTA.

Finally, we need a new mindset. We need to think of the NAFTA region as a single economic unit, and develop a transnational, multidisciplinary vision for approaching and solving our problems. We do not, and probably never did, exist as three countries with three separate and homogeneous styles. Within the U.S. there are large areas that function as developing nations, and within Mexico there are state-of-the-art facilities that control hazardous exposures as well as possible. Certain problems exist throughout the region, and NAFTA provides a vehicle for collaborating to address them. This conference, and the work that will grow out of it, is an excellent example.

References

National Research Council. *Risk Assessment in the Federal Government: Managing the Process* (Washington DC: National Academy Press, 1984).

Frumkin H, Hérnandez Avila M, Espinosa Torres F. Maquiladoras: a case study of a free trade zone. *Int J Occup Env Health* 1995;1:96-109.

Raworth P. Regional harmonization of occupational health rules: the European example. *Am J Law and Med* 1995;21:7-44.

Earth System Science for Sustainable Development of Earth Resources

William S. Fyfe*

ABSTRACT

Our ultimate resources come from the Sun and the Earth System. At this time we face a world crisis in energy, water and soil resources. Global pollution problems are increasing in almost all nations. There is an urgent need to use our vast knowledge with a new wisdom and this requires new integrated teams of experts to provide the best alternatives for the needed new development problems. These teams must include economists, engineers and a broad spectrum of scientists. Today it is possible to produce clean energy, stop soil erosion, clean water and reduce and use wastes of many types.

There is a great need to improve science, natural science, education at all levels. All people must know how our planet works and that fluctuations, some inevitable, some controllable, will occur and that we must prepare for them. And it is clear that protecting the environment, eliminating pollution, preserving biodiversity is good economics. Ecology-Economy cannot be separated.

INTRODUCTION

It is difficult to comprehend the future impact on the earth system the of human population increase expected next century. Barring some vast (but not impossible) catastrophe, our numbers will reach over 10 billion and some, even suggest we must contemplate 20 billion. In the 1995 Worldwatch Report, Brown writes that China, for example, will add a Beijing each year for the next forty years because most people live in cities and urban regions. The situation was well stated recently by Koshland (the former editor of Science) "First of all, it is important to identify the main villain as overpopulation. In the good old days (viewed through the myopia of nostalgia), the water, air, flora and fauna existed in an idyllic utopia. But, in truth, there were famine, starvation, horses and buggies that contributed to pollution, fireplaces that spewed forth soot from burning soft coal, and water contaminated with microorganisms. The humans were so few, and the land so vast, that these insults to nature could be absorbed without serious consequence. That is no longer true."

---

* Department of Earth Sciences, University of Western Ontario, London, Ontario, Canada N6A 5B7

There is a leading question which all educated world citizens must ask. Can we provide adequately for the next 5 billion without destruction of the most fundamental components of the life support system? This century has been truly remarkable. The scientific giants of the early 1900's (Einstein, Rutherford, Planck, Bohr) wrote the rules of modern science and technology, the relations between energy and matter. Their work has stood the test of time remarkably. Their new science led to our present age of observation and information technologies. In large part they contributed to the new medical science which has led to our greater average longevity and the dramatic decline in infant deaths around the world, trends which continue. At the same time, our information systems, in particular television, have led to a new age of expectation for on TV all people are beautiful and live in nice houses and all drive a Buick! In the 40's Aldous Huxley wrote that we have treated nature with greed, violence and incomprehension. There is no question today, that the worship of wealth has never been greater, the gap between rich and poor expanding, while violence spreads across the world with over 40 nations in some form of destructive conflict. But incomprehension, "I don't know what I am doing" is no longer an excuse. Our planet has become small, and our future common, as stressed in the famous report of the World Commission on Environment and Development, (1987), The Bruntland Report.

When we consider the life support system and the future problems faced by Homo sapiens, certain key areas are obvious. They include (in no order of priority)

Energy resources
Water resources
Food resources
Materials
Waste management
Geo-fluctuations, climate fluctuations, and the need for surplus
Biodiversity for security.

The word "sustainable" has become popular. It is a simple concept and requires the answer to a simple question - "will I leave the planet in better condition for all species who support me and follow me?"

In these brief notes I wish to consider the emerging role of the Earth System Sciences in achieving sustainable development and providing the necessary resources for all people. It is interesting to note some recent comments in lead journals "Earth Sciences, job prospects on shaky ground" (Holden, 1994); "Geological programs come under threat" (Macilwain, 1994); "Geology is under attack" (Rossbacher, 1995). Are the earth sciences not needed for future sustainable development (see Fyfe, 1994)? In a recent editorial in Chemical Engineering News (1995), Heylin quotes Stokes of Princeton, "better prospects for a new contract for science lie with common recognition of the importance of what he calls use-inspired basic research. He says basic science and technology move ahead together, each, in turn, inspiring and supporting the other in an endless overlapping process. At times, basic science does indeed trigger new technologies, even new industries. Equally, technological needs inspire basic research".

During the periods of conflict as in the period 1939-45, most scientists became applied scientists. But today, we face a global crisis of far greater magnitude than that of the last world war, the crisis of our survival. We must respond.

THE GEOLOGICAL MAP

As never before, there is need for highly precise geological mapping on scales appropriate to the development problems being considered. Such maps must be precise in describing the timing of events and must be precise in 3 dimensions. For example, if we consider the growth of megacities, the geological knowledge required to prevent-reduce costly engineering mistakes (e.g. Kobe, Japan) to provide and protect water, to prevent pollution etc., is of a detail and range far beyond most present mapping systems. Recent studies, as with the German deep drilling experiment (the KTB), clearly show that our techniques for deep remote sensing are far from adequate today. I was recently on a field trip with an excellent group of Portuguese (Lisbon University) structural geologists. They were concerned with mapping a region of some interest in terms of a site for nuclear waste disposal. The region was well known for some major fault structures. But their detailed studies, clearly revealed the complexity of the stress patterns and showed that micro-fault systems were present with a frequency distribution of tens of metres. This type of detail is essential to the planning of any major engineering project. At a recent meeting in Norway, Swedish geologists reported on the use of good maps in planning the exact location of new highways resulting in large cost reductions. I will return to other aspects of the types of information required in modern maps below with reference to specific development problems.

ENERGY PROBLEMS

At the present time, the bulk of world energy comes from the combustion of coal, oil and gas, all these resources are non-sustainable, natural capital. The thoughtless waste of such valuable resources is a global disaster. Of the fossil carbon sources, only coal and certain types of carbon rich sediments have resources of interest for more than a few decades. In a general way, there has been little change in burning technologies - add air - burn - and exhaust to the atmosphere.

There is no need here to discuss the potential future impacts of the climate changes related to the fact that we have rapidly changed the chemistry of the atmosphere. We normally discuss CO_2, CH_4, and acid compounds. But as we have stressed previously (see Fyfe and Powell, 1995), many coals contain significant quantities of all halogens F, Cl, Br, I and the steadily increasing ozone catastrophe may be influenced by such combustion. Also, time after time, the detailed chemistry of coal and coal ash, is not well known and many coals have significant quantities of elements like uranium and arsenic, and an array of heavy metals immobilized in the reducing, sulfur-rich, medium of coal.

It is certain that nations like China and India will depend on coal for decades to come. Can we change the technology at reasonable cost? Can we reduce the environmental impact of coal combustion? I think the answer is positive. We have been studying the Fixation of CO_2 and organics in the cracked, permeable basalts in the caves of Kauai, Hawaii, deep beneath a very heavy forest cover. Every crack is covered with white stuff (silica, clays, carbonates) formed by the action of organics with the basalt, a process mediated by ubiquitous bacterial biofilms. We now know that bacteria can live to depths of over 4 km, up to 110°C, in favourable locations (Pedersen, 1994). Can we use such processes to fix the exhaust gases of coal combustion? For sure, certain rock types will be better than others and volcanics with Ca-feldspars and other Fe-Mg phases should be ideal as in Hawaii. Recently on a field trip in China, (East of Beijing) we discussed the possibility of using their rapidly exploited oil-gas fields for disposal of wastes of many types. If a basin can isolate oil-gas for millions of years, it surely has capacity to isolate wastes. And generally, oil field structures are well known.

The growing knowledge of the deep biosphere also raises the possibility, with certain types of carbonaceous sediments, of using microorganisms for in situ methane production. In place of opening deep mines with all the related water pollution problems could it be possible to produce bio-gas?

But ultimately, the world must move to solar energy of all types (photovoltaics, wind, tidal) and geothermal energy. There is no shortage of energy sources on this planet. Wind energy use is increasing across the world and photovoltaic devices are becoming more efficient and cheaper (see New Scientist, 1995). Geothermal sources are normally associated with regions of high heat flow (volcanic systems) but for some purposes (city heating, greenhouses and aquaculture systems), the normal geothermal gradient can provide background heating. All such potential use requires exact knowledge of deep geologic structures, porosity, permeability and geochemistry.

At the present time only a small fraction of world energy is provided by hydro-electricity. There are regions of the world where there is still potential with river systems (e.g. parts of Africa, Brazil). But there are problems. River flow, runoff, is necessary to keep the land surface clean, particularly with species like salts. Climate, rainfall, is not constant as we have seen dramatically in current years and in dry periods, runoff may be greatly reduced. We now know that major fluctuations and changes in runoff patterns can greatly perturb the marine biomass and even ocean current patterns which in turn can perturb local and global climates (e.g. the Younger Dryas cold event, see Fyfe, 1993).

WATER

According to Postal (1992), today, forty nations have a crisis of water supply. In many places, uncontrolled extraction of ground water (mining) is being used to promote non-sustainable increase in food production. In vast areas bad water management has led to salinization of soils and massive pollution by agri-chemical residues. I was recently in Calcutta and with their Institute for Man and the Environment, was introduced to their major problem of groundwater arsenic pollution which has led to serious health problems with very large numbers of people who live from ground waters. Where does the arsenic come from? This problem is not solved but arsenic compounds have been used for a long time in rodent control.

In the developing world there are few places where the total water cycle is adequately described. While sea level is slowly rising, in many sensitive coastal regions, land subsidence and apparent sea level rise is associated with subsidence, compaction, following the mining of groundwater. Again, exact geoscience is required to describe the water resource potential of any region and to live with the natural fluctuations in precipitation.

FOOD-FIBRE

At this time, at least one billion humans do not have an adequate supply of food of well balanced nutritional values (Sadik, 1989). Across the world wood is becoming an expensive and declining commodity. Despite the electronic revolution, the use of paper products is increasing (per capita, 3x in the last 40 years). And the world's marine resources are declining at an alarming rate. Again the rich-poor gap is dramatically increasing the nutritional difference in the world's population.

Sustainable food-fibre production depends on climate, climate fluctuations, soil quality and water resources. Knowledge from the geosciences is involved in all these parameters. Given that we are not adequately providing nutrition for the present human population, what are the prospects for the next 5 billion?

All organisms require a large array and balance of the chemical elements (about 50) for efficient production of the organics needed for life. The geochemistry and mineralogy of soils are critical in estimating the capacity of a soil for sustainable organic productivity. According to the Worldwatch Institute, top soil loss globally is approaching 1% per year. The technologies exist now for erosion and salinization control but such technologies are not adequately used. But there is great need for new soil maps which clearly show good soils, soils for forests only, and leave it alone! (Fyfe, 1989).

Given the chemical and physical properties of a soil, additives may greatly enhance bio-productivity. Often such additives require the addition of simple mineral materials containing species like K, Mg, Ca, P.... and appropriate trace metals like Co, Mo, etc. which may be critical in biofunctions like nitrogen fixation. The types of additives useful may be closely linked to soil type and climate. For many situations as with the laterite soils of the humid tropics, slow release, mineral fertilizers (K in feldspars, rock phosphates) may be more effective and less wasteful than soluble chemical fertilizers.

MATERIALS - MINERALS - MINING

Advanced societies use about 20 tonnes of rock per person per year for their needs. Most is for various forms of construction, highways, buildings, etc. Giant mining operations include those for fertilizers, ores like iron, and coal mining. In an experiment we recently did with my research group the average "student" gold ring weighs about 10g and represents the processing of about 3 tonnes of rock. It is amazing but humans in general do not have a clue where their resources come from, the impacts on the ecosphere, geosphere, hydrosphere of mining operations. With modern understanding of Earth convection, our understanding of the resource base and prospecting strategies have dramatically increased. If the world needs more copper, we know where to look and we find it.

But because historically, mining technology has been careless, increasingly many large companies move to developing countries with less stringent environmental laws. For example, I was amazed when the NAFTA agreement was signed, it was clearly stated that local environmental regulations apply. What of the future of the use of mineral resources? For a population of at least 10 billion projected for next century, we must consider the human modification of 100 km3 of rock per year. Most will be ripped from the near surface, but much from increasing depths where mining perturbs the groundwater resources.

First, the world must move to more efficient recycling and for this to be successful requires careful quality control at all stages of the use - mining to fabrication. In mining, there must first be careful 3D mapping of structures, permeability, porosity, faulting, etc. to accurately assess the environmental impacts. There is great need for extreme quality control in the extraction of ore materials. The total geochemistry must be known, the desired and the undesired elements. Such data must be available to plan the mining technology and to assess the environmental impact of the operation. Waste products from mining must be studied for potential uses in construction, soil re-mineralization, etc. The growing knowledge of microorganisms at depths (now over 4 km) opens a host of new technological opportunities. Silica secreting bugs might be used for permeability control. Metal secreting bugs can be used for removal of heavy metals as has been well demonstrated. And a host of new possibilities must be considered for "in situ" metal extraction via sulfide oxidizing species. The same is true for "in situ" methane production from carbonaceous sediments. With all such things there is need for cooperation between geologists, geo-microbiologists, hydrogeologists, engineers and economists. Far too little thought has been given to the end use of mines. In some cases by careful planning, these could become waste disposal sites for urban areas, a growing world problem.

I think there is no doubt, that with the correct team for planning from the start, the environmental impact of extracting resources from the crust can be vastly reduced. And I am sure that in many cases, the long term economics of the operation will be improved. We must also watch for new needs. I was interested to read in the British journal the Economist, that there is a world shortage of high purity silicon for modern electronic devices, and soon we will see vastly increased use of photovoltaics. For such purposes, there is a giant difference between 99.99% SiO_2 and 99.9999% SiO_2.

I have always been intrigued by the possibilities for use of near ocean ridge sites for metal extraction and energy production (cf the Salton Sea Thermal fields). On Canada's well described Juan de Fuca system, by drilling through the impermeable, sediment cover with thermal gradients of up to 300°C/km, one might simultaneously extract metals and energy.

WASTE MANAGEMENT

This century will go down in history as that of careless technologies and waste production. There is no doubt that the long term costs can be staggering (as with the arsenic pollution in India mentioned above). The complexity of modern wastes is enormous, from organics, to radionuclides, urban garbage, etc. And it is amazing that the nuclear industries developed before any serious attention was given to wastes - the we will do it when necessary philosophy. Ontario, Canada, where nuclear electricity dominates the system, now estimates that it will spend at least 15 billion dollars on nuclear waste disposal in the next decades and at this time no "best" site has been proposed. And it is strange that only recently has a combustion gas like CO_2 been considered a waste product. Combustion has changed the concentration of many critical components of the Earth's atmosphere.

I think the time has come to drastically change our philosophy on wastes. First we must precisely describe the nature of waste, chemistry, etc. etc. Then we must search for uses for the waste and recognize that time after time it can be a resource. Denmark recycles

97% of its paper, Canada 17%. And the secret of domestic waste management is the five minutes a day spent in separating the components. As mentioned above, we have shown that by using careful geochemistry, many types of coal ash (but not all) can be a valuable soil additive. The same is true for most urban sewage as long as it is not mixed with other toxic wastes from say the chemical industries. Again we need teams of the appropriate scientists, not just engineers! One case I would like to again emphasize is that of the gas products from combustion, CO2, NOx, SOx, and as we have shown with halogen bearing coals which are common, possible halogen-organics. Are the last partly responsible for the growing problems of stratospheric ozone destruction? As mentioned above, can we dispose of these gases below ground and not simply vent to the atmosphere? I was recently in China in the Beijing region with catastrophic atmospheric pollution from combustion and other industries. The cost reduction on public health of reducing all the lung problems might well cover the additional engineering costs. Japan is seriously considering the marine dumping of CO2.

Every waste product requires unique approaches. Many wastes can be resources, and for most secure disposal is possible. For geological disposal we require a new precision in the total description of the subsurface environment.

GEO-FLUCTUATIONS

There is no need here for lengthy discussion of the problems of the inevitable geo-fluctuations that occur. The Earth environment is not, never was, never will be, constant. Causes of fluctuations, the year without summer, are complex but must be considered in the life support system. They imply that for security there must be surplus and their must be biodiversity. I think one can predict with certainty, that given population growth and present planning systems, the next events like the mega-volcanic eruptions of early last century, will see a vast social catastrophe unless we are prepared to plan now.

CONCLUSION

As world human population continues to grow, to move to over 10 billion next century, the need for exact geoscience must be a priority in planning the needed future development of the support systems. And there is urgent need to improve the communication and effective cooperation between all the experts in modern science and technology, and economists, engineers, politicians and all educated citizens. At this time Europe leads in demonstrations that sound economic and environmental policies are not in conflict but must form a working partnership. We must and can reduce pollution and wastes. we must recognize the limits of the Earth system, we must develop holistic natural science. Given the future numbers of humans on Earth, the cost of errors will become intolerable.

REFERENCES

Brown, L R, 1995. State of the World, 1995. 255 p (W. W. Norton & Co., New York).

Fyfe, W S, 1989. Soil and global change. Episodes, 12: 249-254.

Fyfe, W S, 1993. The life support system in danger: challenge for the earth sciences. Earth Science (Japan), 47, No. 3:179-201.

Fyfe, W S, 1994. The role of Earth Sciences in society. Nature and Resources, UNESCO, 30:4-7

Fyfe, W S and Powell, M A, 1995. Halogens in coal and the ozone hole, Letter to the editor. Chemical Engineering News, (American Chem. Soc.) April 24:6.

Heylin, M, 1995. Science for the 21st century, Chemical Engineering News, (American Chem. Soc.) March 13:5.

Holden, C, 1994. Job prospects on shaky ground. Science, 266: 1316.

Macilwain, C, 1994. Geological programmes come under threat. Nature, 372:715.

New Scientist, 1995. Cheap solar power, January 14:11.

Pedersen, K, 1994. The deep subterranean biosphere. Earth Science Reviews, 34:243-260.

Postal, S, 1992. Last oasis - facing water scarcity. 239 p (W. W. Norton and Co.: New York).

Rossbacher, L A, 1995. Geology is under attack. Geotimes, Feb.:68.

Sadik, N, 1989. The state of world population, 1989. 34 p (United Nations Population Fund, U.N.: New York)..

World Commission on Environment and Development. 1987. Our Common Future. 383 p (Oxford University Press: Oxford)

Air Pollution Engineering:
Source Reduction and Emission Control

James A. Mulholland*

During the late 1940s and 1950s, public awareness of the impact of air pollution on human health was heightened by a series of photochemical smog episodes. In October 1948, 21 people died and 6000 became ill in the small town of Donora, Pennsylvania, from exposure to a mixture of fine particles and sulfur dioxide mist from steel and metal processing plants. In December 1952, 4000 died in a London smog epidemic resulting from residential heating, industrial, and utility emissions. Both events occurred over periods of several days due to unfavorable atmospheric mixing conditions. Photochemical smog was not the only air quality problem resulting from anthropogenic emissions that became apparent during this time. Research in Canada and England in the 1950s and 1960s showed that acid precipitation near industrial regions could be attributed to oxides of nitrogen and sulfur emitted from combustion processes. The magnitude of the acid precipitation problem became clear in the 1970s when reductions in fish populations and forest damage in the mountains of northeastern U.S. and Canada were documented. These environmental tragedies led to the passage of Clean Air Act (CAA) legislation in many industrial countries. Health-based ambient air quality standards (AAQS) were established to identify acceptable levels of six criteria pollutants: ozone, particulate matter, sulfur dioxide, oxides of nitrogen, carbon monoxide, and lead. Source performance standards (SPS) were established to limit stack emissions.

Since passage of air pollution regulations, evidence of several cases of improved urban air quality have been documented. For example, visibility in Los Angeles is much better today than it was twenty years ago. Air quality success stories have occurred due to the implementation of cleaner, more efficient technologies and stack gas cleaning technologies, not because of reduced productivity or lower standard of living. These improvements have been made possible by enormous gains in understanding of the chemical, physical, and biological processes that govern air pollutant generation and destruction, transport and fate, and human health impact.

---

* Assistant Professor, Environmental Eng., Georgia Inst. of Tech., Atlanta, Georgia 30332-0512

Nonetheless, areas remain all over the world in which public health is jeopardized by air pollution. In Mexico City, carbon monoxide, lead, and smog are chronic problems that have been associated with a number of human health outcomes, such as asthma, dysplasia and other nasal cytological effects. The U.S. EPA estimates that 75 million people live in ozone non-attainment areas in the U.S. Moreover, a growing body of medical and epidemiologic data is motivating current discussion of proposals to lower the U.S. ozone standard from 0.12 (1 hr average) to 0.07-0.09 ppb (8 hr average), and to change the particulate matter standard to include only particles less than 2.5 mm instead of those less than 10 mm. In addition to AAQS criteria pollutants, increasing awareness of the harmful effects of trace toxic chemicals in the air, such as certain polycyclic aromatic hydrocarbons (PAH) and vinyl chloride, has resulted in passage of air toxics legislation in the U.S. and elsewhere. Industrial accidents, such as the release of methyl isocyanate from a Union Carbide carbaryl manufacturing plant in Bhopal, India, in 1984, and the release of dioxin from an explosion at a trichlorophenol manufacturing plant in Seveso, Italy, in 1976, have provided dramatic evidence of the effect that air toxic releases can have on human health.

Large scale regional and global air quality degradation also has been observed. Holes in the stratospheric ozone layer have not disappeared despite reduction in the use of chlorofluorocarbon (CFC) aerosols, the source implicated by 1995 Nobel laureates Molina and Rowland in the 1970s. Regarding global warming, NASA figures indicate that 1995 replaced 1990 as the warmest year on record, and that all10 of the warmest years in the past century have occurred in the past 15 years. In November 1995, a report by the United Nations Intergovernment Panel on Climate Change concluded that global warming is no longer only a theoretical threat. Over the next century, global average temperatures are projected to rise between 1 and 3°C. In industrial countries, conservation and greater efficiency could cut the emission of greenhouse gases by up to 30 percent over the next 30 years, but the International Energy Agency estimates that emissions in developing countries will rise over 100 percent by the year 2010.

Continued study and information exchange is needed to address issues of air pollution source assessment and control, atmospheric transport and fate, and human health and environmental impact. The focus of my research is on the formation and control of air pollutants from thermal processes, with emphasis on the development of predictive tools that can be used as input to atmospheric models of pollutant transport and fate as well as health risk models that estimate potential impact. In the rest of this paper, I describe my research interests by summarizing three broad areas of air pollution engineering research in which I have worked; selected publications are cited. I look forward to a discussion of these and other areas as well as collaborative research opportunities with the multidisciplinary and international workshop participants.

1. Reaction Pathways in the Formation of Toxic Combustion Byproducts

Toxic emissions from a well designed and operated combustion device, such as an incinerator, can be negligible. Equipment malfunction or departure from normal operating conditions, however, can lead to combustion byproduct emissions that pose a risk to human health. Of particular concern is the formation and release of PAH and soot, chlorinated organic byproducts such as dioxin, and toxic metal aerosols. The presence of chlorine in incinerator waste streams promotes the formation of all of these. Chlorine increases the volatility of many metals present in wastes, including lead, cadmium, and nickel [Mulholland and Sarofim, 1991]. In addition to the volatilization-condensation pathway of metal aerosol formation, explosive fragmentation routes have been identified. The presence of chlorine also suppresses flame ignition and promotes pyrolytic growth to soot [Mulholland *et al.*, 1992a]. In this process, polychlorinated biphenyls (PCB) are formed from chlorobenzenes [Mulholland *et al.*, 1992b], perchloroaromatics from highly chlorinated industrial solvents such as trichloroethylene [Mulholland *et al.*, 1992c], and highly mutagenic tars from polyvinyl chloride source ethylene dichloride [Mulholland *et al.*, 1994].

By studying chemical speciation and phase transformations that lead to the release of these toxic combustion byproducts, conditions can be identified that avoid their formation. In addition, reaction pathway analysis can be used to predict the types of byproducts produced, and the distribution of isomer and homologue groups. For example, the distribution of PCB isomers produced in the pyrolysis of chlorinated benzenes depends on steric effects that are predicted by molecular orbital modelling [Mulholland *et al.*, 1993]. Similarly, biaryl reactions producing PAH dimers and dioxin can be modeled and product distributions predicted. The application of computational chemistry tools for predicting toxic byproduct distributions will aid the assessment health risk from thermal process emissions by providing a means of estimating the large number of compounds whose exhaust gas concentrations are not measured due to inadequate detection limits or high analytical costs.

2. New Technologies for Reduced Air Emissions

Often the least expensive and most efficient method of reducing source emissions is process modification to prevent air pollutant formation. Thermal and mixing conditions in combustion processes govern the formation of organic byproducts and oxides of nitrogen (NO_x). Simultaneous control of these pollutants is difficult because NO_x formation is favored in high temperature and oxidative environments whereas carbon growth occurs in low temperature and pyrolytic environments. Fuel gasification and staged combustion technologies have been developed to address these issues [Lanier *et al.*, 1986].

When the formation of volatile organic compounds (VOCs) and other air pollutants cannot be avoided, technologies need to be available for destroying or removing them in stack gases. Incineration can destroy those VOCs present in high concentration, but is much less efficient at destroying VOCs present at low levels. Several alternative catalytic and thermal destruction technologies are being developed that are efficient at destroying VOCs present in gas streams at low concentration. Two examples are corona discharge processes and destructive adsorption by particle bed reactors. The latter is being tested in our laboratories, using calcium and magnesium oxide nanoscale particles to destroy various chloro-organics, including carbon tetrachloride and trichloroethylene, and dimethyl methylphosphonate. More work is needed to make the generation of nanoscale surfaces economical and to make partical regeneration more efficient. In the area of NO_x control, thermal de-NO_x chemistry can reduce NO_x emissions by as much as 90 percent in the laboratory, but is often limited in field application by inadequate control of the temperature and mixing environment. Technologies are being developed that widen the temperature window in which NO_x destruction occurs.

3. Air Pollution Source Measurement and Exposure Assessment

Fugitive emissions from a variety of industrial and commercial processes are major sources of air pollution. These sources are difficult to assess because they are disperse and pollutant concentrations are small. For example, U.S. CAA amendment Title V permits often rely on the use of chemical inventory data and mass balance analysis. We are finding from air emission assessment at a textile plant that VOC emissions estimated in this way can be significantly overpredicted in some cases due to thermochemical limitations on VOC release, and significantly underpredicted in other cases due to the emission of compounds not listed on material safety data sheets or produced by chemical reaction. More data are needed to quantify emissions from these sources, and better models need to be developed that relate emissions to the physical and chemical processes.

Another example of emissions that are difficult to assess are the releases of trace amounts of toxic metals and radionuclides from incinerators burning hazardous and low level radioactive wastes. When the amount of these compounds in the waste feed is known, equilibrium analysis of the partitioning of these compounds between chemical species and phases can provide a reasonable estimate of the fine vaporization-condensation aerosol produced at high temperatures [Robinson *et al.*, 1995]. More thermodynamic data are needed, however, on many of the metals and radionuclides of interest. When the waste feed content is not well defined, continuous emission monitors are needed to monitor emissions and ensure safe operation. Further development is needed of instruments that can detect trace species in real time at the levels of interest.

Emissions from vehicles operated under normal conditions are well understood. Emissions from poorly maintained vehicles, and vehicles operated under heavy load (fuel-enriched mode) or during transient conditions (cold start), are not as well understood [LeBlanc *et al.*, 1995]. These emissions contribute significantly to mobile source emissions of carbon monoxide, hydrocarbons, and NO_x. Laboratory research and instrumented vehicle studies are needed to improve mobile source emission models.

Finally, associations between anthropogenic air emissions and human and environmental health outcomes need to be better understood. For example, epidemiologic evidence of associations of asthma and urban air quality needs further exploration, first, to identify particular air pollutants of most concern, and, second, to allocate those compounds to sources. This will require interdisciplinary collaboration ranging from engineers who study source emission and control, to atmospheric scientists specializing in air pollutant transport and fate, and to public health and medical scientists with expertise in environmental health hazards.

References

Lanier, W.S., Mulholland, J.A., and Beard, J.T. (1986). Reburning Thermal and Chemical Processes in a Two-Dimensional Pilot-Scale System, *Twenty-First Symposium (Int.) on Combustion*, The Combustion Institute, Pittsburgh, PA, pp. 1171-1179.

Mulholland, J.A., and Sarofim, A.F. (1991). Mechanisms of Inorganic Particle Formation during Suspension Heating of Simulated Aqueous Wastes, *Environ. Sci. Technol.*, **25**, 268-274.

Mulholland, J.A., Sarofim, A.F., and Beer, J.M. (1992a). Chemical Effects of Fuel Chlorine on the Envelope Flame Ignition of Droplet Streams, *Combust. Sci. Technol.*, **85**, 405-417.

Mulholland, J.A., Sarofim, A.F., Beer, J.M., and Lafleur, A.L. (1992b). Formation of PCBs and Other Biaryls during Pyrolysis of *o*-Dichlorobenzene and Toluene, *Twenty-Fourth Symposium (Int.) on Combustion*, The Combustion Institute, Pittsburgh, PA, pp. 1091-1099.

Mulholland, J.A., Sarofim, A.F., Sosothikul, P., Monchamp, P.A., Plummer, E.F., and Lafleur, A.L. (1992c). Formation of Perchloroaromatics during Trichloroethylene Pyrolysis, *Combust. Flame*, **89**, 103-115.

Mulholland, J.A., Sarofim, A.F., and Rutledge, G.C. (1993). Semiempirical Orbital Estimation of the Relative Stability of PCB Isomers Produced by *o*-Dichlorobenzene

Pyrolysis, *J. Phys. Chem.*, **97**, 6890-6896.

Mulholland, J.A., Sarofim, A.F., Longwell, J.P., Lafleur, A.L., and Thilly, W.G., (1994). Bacterial Mutagenicity of Pyrolysis Tars Produced from Chloro-organic Fuels, *Environ. Health Perspect.*, **102**, Suppl. 1, 283-289.

LeBlanc, D., Meyer, M.D., Saunders, F.M., and Mulholland, J.A. (1995). Carbon Monoxide Emissions from Road Driving: Evidence of Emissions Due to Power Enrichment, *Transportation Research Record*, Transportation Research Board.

THE INTERGOVERNMENTAL MEETING OF HEMISPHERIC ENVIRONMENTAL TECHNICAL EXPERTS 11/1995: PROJECTS AND PRIORITY ISSUES

Luis F. Pumarada-O'Neill*

The 23rd initiative in the plan of action developed at the Summit of the Americas (Miami, 1994) is called Partnership for Pollution Prevention. It builds up on the Agenda 21 meeting and it calls for sound environmental management as an essential element of sustainable development. It calls for a partnership for developing cooperative efforts to develop and improve frameworks for environment protection and mechanisms for implementing and enforcing environmental regulations. This will allow the economic integration of the region to occur in an environmentally sustainable manner. One of the actions proposed was to convene a meeting of technical experts, designated by each interested country, to develop a framework for cooperative partnership, building on existing institutions and networks, and to identify priority projects on specific issues. The Intergovernmental Meeting of Technical Experts was held in San Juan, Puerto Rico on November 6-8, 1995. It was preceded by Advisors Pre-meeting on November 2-3. They were hosted by the government of the Commonwealth of Puerto Rico and sponsored by the Organization of American States, the Pan American Health Organization, and the U.S. Environmental Protection Agency. Its purpose was to design the Partnership for Pollution Prevention and to set an agenda for follow up activities, including reports to the Summit Implementation Review Group (SIRG) and the Sustainable Development Summit in Bolivia.

The Meeting of Technical Experts included panel discussions on Financing Strategies and Generic Environmental Strategies, and Concurrent Working Groups Sessions on:

Lead Risk Reduction
Pesticide Management

* Director, CoHemis, Center for Hemispheric Cooperation in Research and Education in Engineering and Applied Science, University of Puerto Rico, Mayagüez, PUERTO RICO 00681-5000

Water Quality
Sustainable Tourism
Framework for Continuing Partnership

The working group objectives were:

Recommendations to Plenary on Priority Projects
Plan for Project Development and Financing Options
Identification of Interest Countries for Regional Projects
Network of Experts for Information Exchange

The Advisors Pre-meeting provided a preliminary list of priority issues and projects for the workshop sessions. The latter would be cooperative projects that will enjoy the support of national and local governments, non-governmental organizations, and private industries.

The participants included organizations such as Inter-American Development Bank, U.S. Agency for International Development, World Bank, and Environmental Financial Advisory Board. Some of the participating countries included: Venezuela, Barbados, Brazil, Costa Rica, Haiti, Dominican Republic, Mexico, Belize, Ecuador, Puerto Rico, Bolivia, USA, Peru, Chile, Argentina, Nicaragua, Guatemala, Uruguay, and Ecuador.

MEETING OVERVIEW

The meeting of Technical Experts successfully developed a framework for advancing the Western Hemisphere Partnership for Pollution Prevention and identified specific projects that address priority issues. For three day environmental professionals representing 25 government agencies, private-sector associations, and non-governmental organizations labored on these recommendations. They were joined by representation of the sponsoring organizations:

the Organization of America States
the Pan American Health Organization
the U.S. Environmental Protection Agency

and by representatives of international organizations and multilateral banks:

the United Nations Environment Programme
the Organization of Eastern Caribbean States
the Inter-American Development Bank
the World Bank

A key to the success of the Technical Experts Meeting was a Pre-Meeting of Advisors. Also drawn from throughout the hemisphere, the advisors meet for three days to develop a set of recommendations for the technical experts to consider. A transcript of these recommendations was given to the technical experts on their arrival. Keys to the success of both were thoughtful, pre-arranged presentations by advisors, not only on potential projects in the key areas of lead risk reduction, pesticide management, sustainable tourism and water quality management, but also in important cross-cutting areas such as pollution prevention, public-private cooperation, public participation and legislative and regulatory reform. Also very welcome were presentation from the multilateral lending institutions outlining their assistance programs. Keynote remarks by the Governor of Puerto Rico, Dr. Pedro Rossello, stressed the importance of hemisphere development and its consistency with environmental goals. He welcomed the efforts of the technical experts to integrate both goals and pledged the support of his government. Perhaps the greatest challenge of the participants was selecting recommendations from among a broad array of potentially worthwhile projects that broadly addressed the organizations of the Partnership for Pollution Prevention or and its areas of specific concern: lead risk reduction, pesticide management, sustainable tourism, and water quality management. The specific recommendations involve a wide range of types of activities. Some involve education, either of the public or of professionals. Others focus on the development of new technology. Still others involve improvements to the legal system or other development of important policy tools. As inventory of health efforts was selected by the one work group, and two groups, sustainable tourism and water quality management, identified some very defined projects, some with clear geographical applications.

SOME PRIORITY ISSUES AND PROJECTS

The following were extracted from the recommendations of the Advisors, since the final version of the recommendations of the Technical Experts are not yet available.

Water Quality Working Group

The water pollution problems identified in the region include: leaking underground storage tanks, leaching of toxics from landfills, mining operations, agriculture operations, industrial operations, gasoline stations, and salt water intrusion.

The group established the following project objectives:

1. Increase water reuse activities throughout the Latin America and Caribbean (LAC) region, focusing on reduction of fresh water consumption per capita, improving the per capita usage, improving water quality, and improving overall water management through education, technology transfer, training, data collection, and operation and maintenance.
2. Developing an economic framework for environmentally sound water management.

3. Holistic management water programs, primarily for dealing water basins or water sheds with international impacts.
4. Reduction of losses in water supply systems, particularly from bad maintenance: improving the LAC water infrastructure through training, education, technology transfer, technical assistance and others.
5. Reduce the incidence of illness in LAC attributable to a contaminated water supply.

This group proposed the following projects:

- Development of environmental standards; adapt standards for local conditions. Devise a more realistic set of standards that people can work with.
- Cleaner production, using public/private cooperation to develop training, economic incentives and internal audits. Study the different sectors in terms of water usage, pollution prevention, and recycling
- Bringing ownership of water supply and treatment systems to the smaller communities and the communities that are more remote, e.g.: through low cost, low maintenance options for portable water treatment and waste water treatment in the LAC region, such as augmented or constructed wetlands, overland treatments, anaerobic treatments. Begin by making an inventory of available technologies that are that may be adaptable to LAC countries and holding workshops to discuss alternatives with the local people.
- Low salinity waste water or water usage initiatives, such as there exist in the Middle East, e.g.: low salinity usage for crop irrigation or household use.
- Sustainable Disinfection; alternatives to chlorine that can used in smaller communities, such as chlorine dioxide, ozone ionization, or ultra violet light application. Promote the use of the OAS' Inter-American Dialogue on Water Management Network for hydrologists to exchange ideas, information, technology, and concepts.

Pesticide Management Working Group

This group suggested as a main objective related to the use of pesticides and fertilizers the reduction of risks for human health and the environment associated to the use of pesticides and fertilizers. This will require knowledge of alternatives to the use of pesticides and fertilizers, such as products with less toxicity, biotechnology, and changes in agricultural techniques. It may also require institutional capacity strengthening, monitoring and registration measures. This problem needs to be treated within a broad, integrated program.

This group proposed the following projects:

- Develop, implement, and disseminate products and other alternatives which carry less risk than the pesticides in use today. Develop new, integrated management techniques for pest control (Integrated Pest Management).
- Develop programs that promote biological pest control.
- Develop programs that promote cultural changes that may reduce pests and the need for using fertilizer.
- Reduce the frequency of incidents and accidents which involve pesticides. Reduce damages to nature and poisonings which still occur.
- Provide incentive programs for the above.

Sustainable Tourism Working Group

This group proposed the following projects:

- Development of a Marine Park in the Greater Caribbean: design a series of pilot projects for ecotourism in the greater Caribbean area which will utilize marine parks, all of which will be protected areas on a sustainable basis, while providing protection for their resources and attending needs of the local population.
- Galapagos Islands Project: Create and test a model for sustainable tourism which helps to preserve the biodiversity of Galapagos and can be transferred to other areas of the planet for the same objectives. These objectives would be: to preserve the biodiversity; develop methodologies for determining the carrying capacity of the tourist sites and the ecosystem; work out a model management plan for a sustainable and profit-making tourism industry which allows the preservation of the islands' biodiversity.

Working Group on Lead Risk Reduction.

This group proposed the following projects:

- Develop national plans for the phase-out of lead in gasoline. These plans must eventually become appropriate legal instruments such as presidential decrees or national laws.
- Study the presence of lead in the environment of each different country; and make an inventory of the uses given to lead and the health risks and environmental impacts associated with it.

Working Group on a Framework for Continuing Partnership

This group provided the following suggestions at the national level:

- Improve the coordination among existing institutions and/or activities to develop a management plan and identify priorities.
- Maintain institutional continuity in order to encourage, facilitate and monitor actions to implement the plan of action of the Summit of the Americas.
- Develop strengthening systems at the national level for environmental policy, legislation, regulations, compliance and enforcement, public participation and other Summit commitments. Encourage pollution prevention as a means of environmental management.
- Develop means to disseminate, share information knowledge and technology, develop national financing mechanisms and prepare quality projects.

At the international level, the group provided the following suggestions:

- Establish coordinating units at an existing inter-governmental organizations.
- Create and improve networks.
- Have the communication technology available.
- Enhance means to disseminate and share information, knowledge and technology. Serve as a broker among national, technical and financial institutions and networks to improve coordination and communication mechanisms for cooperation among government agencies.

WASTE DISPOSAL: SOME ITEMS FOR DISCUSSION

R. Kerry Rowe*

Introduction

The generation and disposal of waste has become a concern to both the public and to governments. Numerous options are often proposed including the three R's (Reduce, Reuse, Recycle), incineration (energy from waste), composting, bioconversion and landfilling. In North America the primary focus has been on recycling, incineration and landfilling. While these options may all have an important role to play as part of a waste management strategy, there are also some questions that need to be considered; this paper is intended to raise some of these issues.

Life Cycle Assessment and Full Cost Accounting

In principle, any assessment of the merits of different techniques for minimizing and disposing of waste should be based on a life cycle analysis and full cost accounting. According to the Canadian Standards Association Guideline Z760 definition,

---

* Department of Civil Engineering, University of Western Ontario, London, Ontario, Canada N6A 5B9

> Life Cycle Assessment is a tool for identifying the environmental releases and evaluating the associated impacts caused by a product, process or activity. An assessment involves defining the product, process or activity; establishing fully the context in which the assessment is being made; identifying the life cycle stages being covered; evaluating and quantifying, among other things, the energy, water, and materials usage, and the environmental releases at each stage; determining the aggregate and specific impacts of the releases; and developing opportunities to effect environmental improvements.

For example, while it may be regarded as self-evident that a reduction in waste generation can only have a positive net effect, it is not necessarily obvious that re-use and recycling have a similar positive net effect. In order to assess this one must consider the life cycle and perform a full cost accounting that includes consideration of the cost and environmental effects that derive from the re-use or recycling process. In some cases, the benefits will be clear but caution is required not to assume that this will indeed be the case.

It can be argued that concepts of life cycle analysis and full cost accounting should also be extended to other methods of waste disposal (e.g. composting, incineration, bioconversion and landfilling).

Life Cycle Assessment and full cost accounting are in the early stages of development. The International Organization for Standards' (ISO) Environmental Management Committee, TC207, is presently working on the development of an international standard for carrying out Life Cycle Analysis. This type of approach will be important to developing fair trade practices across North America provided it is applied in an equitable manner in all jurisdictions.

Composting

The objective of composting is to biologically stabilize suitable organic waste to the point where the decomposed product can be safely returned to the earth. On the home scale, composting may be regarded as a relatively benign and easy means of reducing the volume of waste requiring other forms of disposal. On the larger scale, even composting has its problems, many of which are similar to those associated with landfills. These problems include issues of controlling leachate generation, odor, dust, noise and traffic impact. Assuming that all these issues can be adequately dealt with, the product of the composting process must either be returned to the ground or disposed of in a landfill. The latter defeats much of the purpose of composting but in order to implement the former the compost must, in many areas, meet certain standards. This is far from a trivial undertaking and, as noted by MacKinnon, in some jurisdictions the compost quality standards for heavy metals such as cadmium, chromium, copper, molybdenum and zinc are such that much of the compost generated is unacceptable for use relating to food and agriculture and hence finds its way back to a landfill. Thus the conflicting desires to return organic matter back to nature and the desire to have very strict regulations relating to compost quality require discussion.

Landfilling

Public concern regarding landfilling can be traced to Love Canal and similar examples of waste disposal where little, if any, thought appears to have been given to the potential impacts. Love Canal was originally excavated in the 1890s as part of a proposed hydroelectric development in Niagara Falls. The project failed and construction was halted with only about 900 m of the originally proposed 10 km canal having been completed. Between 1942 and 1953 approximately 22,000 tons of chemical waste was buried in Love Canal - including large quantities of chlorinated hydrocarbons. Up to the 1970s relatively little attention was given to the potential for subsequent contaminant transport. However, starting in about 1976, chemical odors, contaminant migration into basements and into an adjacent landfill and reported discomfort and illness of residents in the area, resulted in substantial publicity and subsequent remedial investigation (Cohen et al., 1987). However, while it was "Love Canal" that gained public attention, there were in fact three other areas relatively close by where another approximate 343,000 tons of chemical waste had been buried between about 1942 and the mid 1970s and thus the chemical waste in Love Canal itself only represented about 6% of the total chemical waste (including brine sludge) buried in a relatively small region.

The primary sources for groundwater contaminants are waste disposal sites such as landfills, lagoons, injection wells, tailings and leaks or spills (e.g. chemical plants, dry cleaners, manufacturing plants and gasoline stations). Numerous problems have arisen from contaminant migration (both in vapor and liquid forms) through soil. Increased public awareness of environmental problems have resulted in the development of an extensive set of regulatory policies and/or "guidelines" that must be followed both to deal with the clean up of existing problems and to promote operations that will minimize future problems. The investigation, clean up and prevention of contamination has become the primary area of activity of many (if not the majority) of geoenvironmental engineers and hydrogeologists in North America.

Part of the proposed solution to the problems associated with landfills has been to separate waste (e.g. separate municipal solid waste from hazardous waste) and to minimize the disposal and generation of liquid waste. Under the U.S. Resource

Conservation and Recovery Act (RCRA) Subtitles C and D, hazardous and non-hazardous landfills will typically (but not necessarily) have composite liner systems consisting of a geomembrane overlying a compacted 'clay' layer which has a hydraulic conductivity of 10^{-7} cm/s (or less). If properly designed and constructed, these composite liners provide a good potential barrier to advective flow (i.e. the movement of liquids) however the potential for contaminant migration due to molecular diffusion has generally not been well recognized (Rowe et al., 1995). Furthermore, while these proposals represent an advance on the situation where there is no regulation, they may result in gross overdesign in some situations and gross underdesign in other circumstances (see Estrin & Rowe, 1995). The combination of a failure to consider the potential impacts due to molecular diffusion together with the failure to consider potential impacts beyond (typically) 30 years post closure creates a situation where it can be much cheaper and easier to have a landfill approved in the U.S.A. than in Canada.

In contrast to the U.S. approach, in Ontario, Canada, a proponent of a new landfill is required to demonstrate that the landfill will have negligible impact on groundwater (based on what is known and known not to be known) or surface water without any period of limitation. Furthermore, the impacts associated with both molecular diffusion and advection generally must be considered together with the likely service life of the engineered components of the system (MOEE, 1994a,b). While this approach does not guarantee that a facility will not, at some future time, have an unacceptable impact, it does avoid the situation where a landfill would be approved and constructed when, given what we know today, there is a high probability that it will cause an unacceptable impact. The cost associated with strict environmental protection regulation in Ontario can be seen as one of the factors contributing to the significant cross-border transportation of waste.

Controlling Infiltration into Landfills

A second important discussion point that arises out of the Ontario and U.S. EPA approaches is the attitude to infiltration through the landfill cover. For example, under RCRA Subtitle D, a landfill proponent is required to have a low permeability cover as soon as possible, so as to minimize the generation of leachate. This approach has the benefits of minimizing both the amount of leachate that must be collected and treated, and the mounding of leachate within the landfill; it also has the disadvantage of extending the contaminating life-span.

Because of the heterogeneous nature of waste, some leachate will be generated almost immediately, even for landfills designed with low infiltration covers. With low infiltration, it may take from decades to centuries before the field capacity of the waste is reached and full leachate generation occurs. This means that the full capacity of the leachate collection system may not be required for many decades after construction.

However, during this period of time, degradation and biological clogging of the leachate collection system can be occurring unless the waste has been pre-treated to convert it to an essentially inert form. Furthermore, due to the variable nature of leachate generation, it may be difficult to assess whether there has been a failure of the leachate collection system, by monitoring the volumes of leachate extracted.

An alternative philosophy is to allow as much infiltration as would practically occur. In humid climates, this would bring the landfill to field capacity quickly and allow the removal of a large proportion of contaminants (by the leachate collection system) during the period when the leachate collection system is most effective and is being carefully monitored (e.g. during landfill construction and, say, 30 years after closure). The disadvantages of this approach are two-fold. Firstly, larger volumes of leachate must be treated; this has economic consequences for the proponent. Secondly, if the leachate collection system fails, a high infiltration will result in significant leachate mounding.

The low infiltration philosophy is readily suited to meet environmental regulations that only require a limited period during which the landfill must not cause an unacceptable impact (be it 30 years or 100 years). However, engineers have a moral responsibility also to consider the longer-term consequences. In some areas (e.g. Ontario, Canada), there is also a regulatory responsibility to consider environmental protection in perpetuity.

When long-term protection is considered, it may be desirable to find a balance between the low/high infiltration philosophies. Although it is not always practical (e.g. because of after-use requirements), one option is to allow high infiltration during construction of the entire landfill (not just the cell) and for a period after landfilling ceases. Once the landfill has been largely stabilized or the leachate collection system starts to degrade noticeably, then a low infiltration cover would be constructed. This approach rapidly brings the landfill to field capacity and removes a substantial portion of the potential contaminants early in the life of the landfill when the engineered components are at their most reliable, thereby reducing the contaminating life-span. This approach also reduces the infiltration and potential mounding of leachate after the low infiltration cover has been constructed, and hence minimizes potential problems once the performance of key components of the engineered system (like the primary leachate collection system) have degraded.

These factors should be considered in selecting the barrier system, in developing the monitoring and contingency plans, and in assessing the service life and contaminating life-span of the facility.

It should also be recognized that while it may be possible to construct relatively tight landfill covers that permit very little infiltration into the landfill, these covers will degrade with time and will require perpetual care. The likelihood of this care being provided and costs associated with perpetual care require very careful consideration.

The Landfill as a Bioreactor

Until relatively modern times, 'landfills' caused little concern for groundwater because the volume of waste being disposed of was small relative to the surrounding environments' capacity to assimilate the waste products. However, modern urbanization has brought with it increased concentration of waste (i.e. larger and larger masses of contaminant being disposed of in smaller and smaller areas so as to save space) as well as new, man made chemicals many of which are more toxic and are not as readily biodegraded as natural substances. This in turn has caused problems such as that at Love Canal as discussed previously, and requires much higher levels of natural protection and/or engineering to minimize impacts on water resources. There is considerable potential for using landfills as bioreactors which can break down waste. To do so safely requires either a natural or man made barrier system to retain the leachate during the degradation process, a supply of fluid and a design which encourages breakdown of the waste and the organic constituents in the leachate. Since not all chemicals will be degraded and some could build up to levels that may inhibit degradation, it is also necessary to have some mechanism to remove leachate.

Modern landfills typically have leachate collection systems which are primarily intended to control the leachate mound on any barrier system. One of the problems associated with these systems is the potential for clogging (e.g. see Brune et al., 1991; Rowe et al., 1995). However, the processes that give rise to clogging are the same processes that give rise to a breakdown of organic compounds in landfills and hence with appropriate design, there is the potential to have a "leachate collection" system which also serves as a fixed film bioreactor which can 'clean up' the leachate before it is collected for treatment. There is evidence to suggest that not only conventional constituents of leachate (e.g. volatile fatty acids) but also chlorinated organic compounds can be broken down (e.g. see Rowe, 1995; Kromann et al., 1995; Johansson et al., 1995).

Past poor practice has resulted in a move away from co-disposal of municipal and hazardous waste in North America. However, in the U.K., many advocate that co-disposal can be a responsible approach for dealing with hazardous waste. This issue warrants re-examination in North America.

Energy From Waste

The burning of waste has a long and tarnished history. Modern incinerators have the potential of recovering considerable amounts of energy while controlling gas emissions to very low levels. However, the facilities are also expensive and require considerable care and maintenance to ensure that gas emissions are kept at accept-able levels. In Germany, there is a significant move towards burning that waste which can not be dealt with by the three R's and composting. The arguments in support of incineration are the generation of energy and the stabilization of the waste. Both aspects need to be carefully considered. In particular, while incineration decreases the volume of waste and eliminates most organic constituents, it can also result in a waste that has much higher inorganic concentrations (e.g. heavy metals) and hence the design of a facility for safe long-term disposal of these wastes is far from trivial. The tradeoffs need to be carefully considered.

References

Brune, M., Ramke, H.G., Collins, H.J. and Hanert, H.H., "Incrustation processes in drainage systems of sanitary landfills", Proceedings Third International Landfill Symposium", Cagliari, Italy, 1991, pp. 999-1035.

Cohen, R.M., Rabold, R.R., Faust, C.R., Rumbaugh III, J.O. and Bridge, J.R. "Investigation and hydraulic containment of chemical migration: Four landfills in Niagara Falls," Civil Engineering Practice, 1987, pp. 33-58.

Estrin, D. and Rowe, R.K., "Landfill design and the regulatory system", proceedings Fifth International Landfill Symposium, Sardinia, Italy, Vol. 3, 1995, pp. 15-26.

Johansson, E., Ejlertsson, J., Karlsson, A., Orlygsson, J. and Srensson, B.H., "Anaerobic degradation of perchloroethylene to vinyl chloride by microorganism in waste", Proceedings, 5th International Landfill Symposium, Sardinia 95, Vol. 1, 1995, pp. 143-148.

Kromann, A., Ludvigsen, L., Christensen, T.H., "Degradability of chlorinated organic compounds in landfills", Proc. 5th International Landfill Symposium, Sardinia 95, Vol. 1, 1995, pp. 135-142.

MacKinnon, W. "Large scale composting - Unreasonable standards will prevent waste diversion goals," Hazardous Waste Materials, 1995.

Rowe, R.K., "Leachate characterization for MSW landfills", Proceedings Fifth International Landfill Symposium, Sardinia, Italy, Vol. 2, 1995, pp. 327-344.

Rowe, R.K., Hrapovic, L. and Kosaric, N., "Diffusion of chloride and dichloromethane through an HDPE geomembrane", Geosynthetics International, Vol. 2, No. 3, 1995a, pp. 507-536.

Rowe, R.K., Fleming, I., Cullimore, D.R., Kosaric, N., Quigley, R.M., "A research study of clogging and encrustation in leachate collection systems at municipal solid waste landfills", Submitted to Interim Waste Authority, 1995b, 131 pages.

SUSTAINABILITY, RISK AND DECISIONS

John Shortreed*

INTRODUCTION

Sustainability is defined here as an objective to maximize the potential for a quality life for present and future generations. It is proposed that this be measured by the UN Human Development Index (HDI) (UNDP, 1995). A condition that the HDI remains constant or is increasing is taken as confirmation of sustainability.

Risk, the possibility of a loss, is always present. Individuals, organizations and countries must consider risk every time a decision is made. Risk along with costs and benefits of alternative choices are "balanced" by the decision maker in making a decision. The situation, environment, and circumstances under which the decisions are made are referred to here as the risk decision framework. The risk decision framework includes sustainability.

For sustainability the losses of interest are loss of health, loss of economic and social
development, and loss of environmental quality.

Environmental quality is considered to have two aspects, aesthetics and health. Environmental health is measured in HDI by life expectancy. Aesthetics are considered as benefits that contribute to the quality of life. In general the aesthetics component of
environmental quality can be considered to be measured by the "income" available to
individuals measured in purchasing power parity (PPP) expressed as GDP (gross domestic product) per capita in the HDI.

Other dimensions of environmental quality; species diversification, food production capability, and natural environments are not measured directly by HDI but may be

---

* Institute for Risk Research, University of Waterloo, WATERLOO, ONTARIO, CANADA N2L 361

measured indirectly by GDP. Although the HDI provides a simple first cut measure of sustainability other measures should also be used.

It is not possible to speak about sustainability independent of population growth. Any increase in population will limit the probability of maintaining sustainability as measured by HDI. Society will jointly choose a rate of population growth and sustainability. One possible strategy would be to limit population growth to at least maintain existing levels of HDI. The choice of population growth is clearly a value judgment. Figure 1 illustrates the relationship between GDP per caput and life expectancy for data from 1985-1988. There is a rapid increase in life expectancy up until $5,000 and then a sharply diminished effect of income on life expectancy. This effect is reflected in the HDI. Figure 1 also illustrates that there is a strong correlation between income and life expectancy.

HUMAN DEVELOPMENT INDEX

The UN Human Development Index (HDI) has been modified extensively since it was first introduced in order to address concerns raised. However, the ratings of individual countries have not changed significantly. Moreover, the ranking of countries by HDI is similar to other indices such as the Life Product Index developed by Lind from the principles of measuring both the length of life and the quality of life as the basis for risk management in the Public Interest (Lind, 1992).

The most recent formulation of HDI includes "A major refinement introduced in 1994, when goal posts were fixed for each indicator to allow analysis over time." (UNDP, 1995) This refinement as well as other modifications to improve the index make it suitable as a candidate for evaluating risk decisions, including decisions on environmental risks.

Table 1 gives the basis for calculating HDI for Canada, USA, and Mexico for 1992. The index is an average of three separate indices:

1. A measure of life expectancy expressed as an index with the goal life expectancy assigned an index of 1.0,

2. A measure of literacy (2/3 weight) and educational enrollment (1/3 weight) again measured against a goal of 100% literacy and 100% enrollment in primary, secondary and tertiary school levels being assigned an index of 1.0, and

3. A measure of income per person using actual PPP income up to $5,000 and then discounting income above that figure dramatically in recognition of the relationship illustrated in Figure 1. Again an index of 1.0 has been set as a goal level of income.

In Table 1 the discounting of income between Canada and the US is at an approximate rate of 0.005 adjusted $ per actual $. Between Mexico and Canada the discount rate is 0.012 adjusted $/$. There is a difference in the adjustment parameters for converting real GDP to HDI "income" to measure life quality and sustainability.

Considering only income and life expectancy in the HDI it is instructive to compare the trade off between life expectancy and real income. A risk that represents 0.4 years of life expectancy is selected, in Canada, this would be equivalent to risks from motor vehicles (Shortreed, 1992). For the data in Table 1 this risk is equivalent, in the sense of maintaining the HDI at the same value, to about $6,500 of real GDP per capita for Canada and $1,800 for Mexico.

Table 1 Human Development Index for Canada, US and Mexico, 1992

	Life Expectancy (years)	Gross Domestic Product $PPP/caput	Adjusted GDP (Fig. 1)	Adult literacy (%)	Education (%) Enrollment	Life Expectancy Index (goal=1)	GDP Index goal=1	Education Index (goal=1)	HDI (Average of Life&GDP& Educ)
Canada	77.4	23,760	5,359	99.0	100.0	.87	.98	.99	.950
US	76.0	20,520	5,374	99.0	95.0	.85	.99	.98	.937
Mexico	70.8	7,300	5,213	88.6	65.0	.76	.96	.81	.842

HDI - UN Human Development Index Source (UNDP, 1995)

There is still some uncertainty about the treatment of income in HDI. For example, applying the previous UNDP definition of HDI income, (i.e. index based on the log(10) of real income), the trade off between a 0.4 year change in life expectancy in 1992 would have been $1,300 real GDP per capita for Canada and $300 for Mexico. Similarly, the Life Product of Lind (Lind, 1992) uses income to the power 1/6 and the equivalent tradeoff would be $750 for Canada and $200 for Mexico. There is considerable uncertainty in the absolute value of the trade off (i.e. one order of magnitude), but there is consistency in the relative values of the trade offs, with the ratios being within 25% of each other.

For the moment the third component of the HDI - literacy and education - will be set aside with the other difficult issues of preservation of natural environments, species diversification, etc., to be considered later in the risk management process.

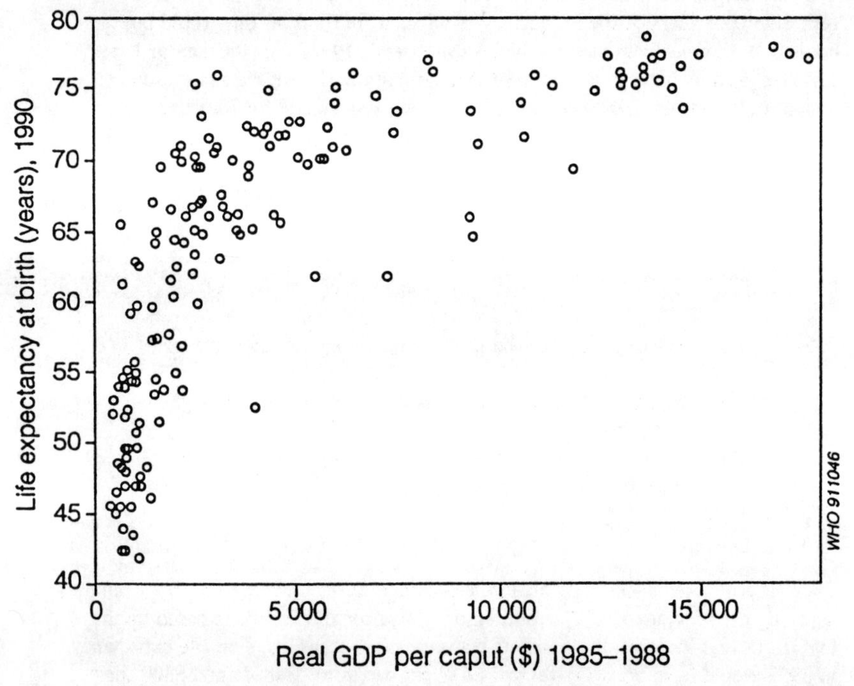

Sources: The figures for real GDP per caput for 1985–88 come from the United Nations International Comparison Project which has developed these on an internationally comparable scale using purchasing power parities instead of exchange rates as conversion factors, and expressed in international dollars. These figures and the figures for life expectancy at birth (1990) were drawn from UNDP, *Human development report 1991*, Oxford, Oxford University Press, 1991.

Figure 1 The Relationship between Economic Development and Life-Expectancy

CRITERIA FOR ENVIRONMENTAL RISK DECISIONS

If sustainability is the goal and if this is appropriately measured by HDI then it follows that the criteria for making decisions on environmental risks will differ between countries. This difference arises from the relative impacts of income and life expectancy and in particular differences in the trade off between income and life expectancy between countries.

The criteria proposed is a HDI performance based criteria. The protection of air quality, or the protection of water from pollution should achieve the same level of protection of life quality in each country.

If a particular environmental threat has the same health effect in two countries then following the HDI measure of life quality, it is correct to spend more money for protection in the higher income country than in the other. This is because in terms of the overall goal of maintaining life quality to do otherwise would lead to a reduction in life quality.

To illustrate the idea, consider a threat that would lead to a loss of life expectancy of 0.04 (i.e. 1/10 of the earlier example or a risk similar to residential fires). Suppose that the usual 80/20 rule applies (80% protection for 20% of the cost of full protection) and that Canada decides to purchase full protection at a cost of $650 per capita. The effect on HDI is neutral, the reduction in income is exactly balanced by the increase in life expectancy (under the assumptions of the current HDI - for other possible definitions there would be a loss of life quality).

For this same situation, suppose Mexico considers two decisions, full protection or 80% protection (at 20% of the cost). Under full protection the HDI would fall from .842 to .841, but under 80 % protection the HDI would remain at .842. Expressing the HDI in terms of life expectancy; the threat is a loss of 0.04 in life expectancy; the decision result is 0 change in life expectancy for Canada(100% protection); for Mexico with 100% protection there is a loss of -0.1 years of life expectancy (greater than the original threat!); and for Mexico with 80% protection there is a gain of +0.004 years of life expectancy.

Similar performance based criteria are currently in effect in Canada and result in different levels of protection for the population. For example, (Paustenbach, 1995) notes:

> *"In most instances, the Canadian government has not considered using cancer potency factors and reference doses developed by the EPA because they see these figures as overly conservative. In general, risk-assessment decisions in Canada are performed on a case-by-case basis. In each case scientists use a weight-of-evidence approach (rather than strict regulatory policy)"*

In other words, there is explicit recognition that there should be different standards in different countries for levels of protection of public health because of different situations.

It must be emphasized that the proposed performance based criteria leads to equivalent human life quality impacts in each country, a very desirable outcome. The myth that this equivalency in human terms can be attained by equal concentration criteria and standardized protection requirements has considerable appeal until the issue is considered using HDI or other composite measures of life quality.

The proposed performance based HDI criteria is only appropriate for risks and other life quality effects that are internal to a country. If the risk was associated with international air transport then the risks are not internal to one country but go between countries and it is necessary to have a world wide set of safety criteria. This is in fact the case for airline safety. For environmental protection of transborder flows of air or water the issues would not be internal to one country and again it would be expected that some common protection level would be considered. But for wholly internal environmental issues there are differences in protection criteria between countries to achieve the best possible life quality.

The use of HDI raises one other issue for environmental criteria - the opportunity safety cost of expenditures. This issue is most clearly seen in the Canadian data in Figure 2, which shows the increase in life expectancy for males in four Canadian cities from 1978 to 1987 (Siddall, 1989). The increase in life expectancy is well in excess of that expected from increasing income levels. Increases in life expectancy have taken place continuously since the time of the industrial revolution, much of it can be explained by the increase in GDP but there is still some unexplained contribution which may be due to the development of health care technology, the improvement in occupational health and safety, or improvements in consumer products such as cars. As shown in Figure 2 there is also considerable differences between cities in Canada, the reasons for this are not known.

The diversion of resources from investment in health, workplace safety, or consumer products may act to reduce the continuous improvement in life expectancy. There is an opportunity cost of expenditure on risk reduction that must be considered when decisions involving environmental risk are made. Unfortunately, the magnitude of this opportunity cost are not known, except that the effect is substantial. A relevant question is why research into the cause of these large increases in life expectancy are not undertaken? The direct trade off between income and life rather than the indirect trade off inherent in the HDI was described in Walden (1854), when Thoreau stated:

> *"the cost of a thing is the amount of what I will call life wich is required to be exchanged for it, immediately or in the long run".*

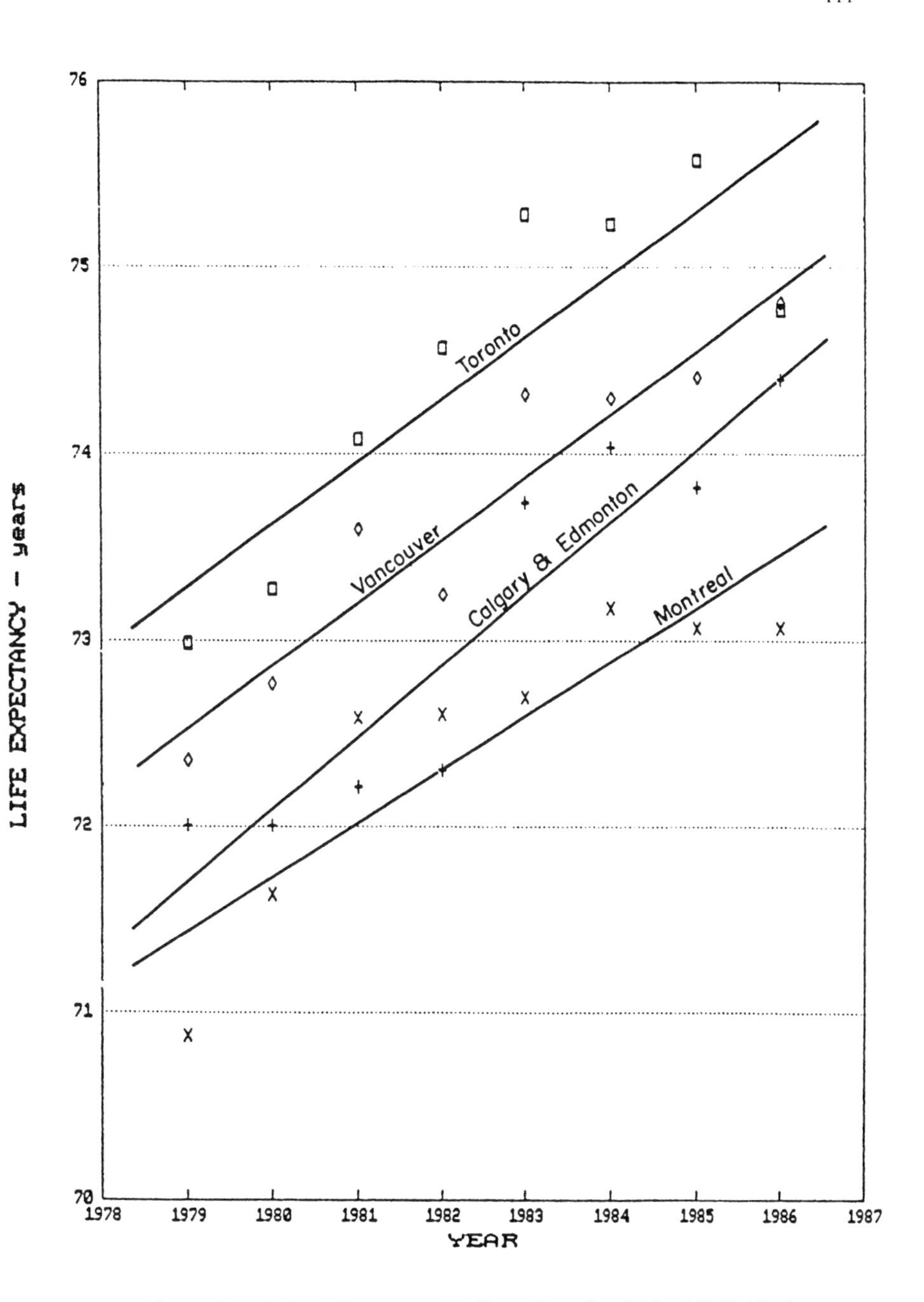

Figure 2 Male Life Expectancy for Four Canadian Cities (1978-1987)

Costs are expenditures of lives through human labor. The costs of risk management controls are the labor expenditure of people working now, in the future, or in the past. Environmental risk management is really a life-life trade off. It is the investment of lives to save life. While it is convenient to use money for costs it must be remembered that it is lives that are being spent. It is inappropriate to use methods that use a "value of life" - the value of a life can be considered to be infinite. However, it is appropriate to speak of the "cost of saving a life" since this indicates the expenditure of lives necessary to save a life and is a life-life comparison that can be used for evaluating risk reduction programs.

Environmental risk management programs should at least "break even". The life-cost of saving a life should be less than the life saved, the ratio should be less than 1.0. If it is more than 1.0, then lives are being wasted. The break even point can be operationalized as the marginal cost of saving a life or the opportunity safety investment. For example, in North America there are unfunded programs in health care and road safety that would save a life for a cost of $CN 300,000 (+or- 100%) (JCHS, 1994). This view of the value of income in terms of life is stronger than that implied by the proposed HDI performance criteria.

The joint committee on Health and Safety, of the Royal Society of Canada and the Canadian Academy of Engineering have adopted the following three principles for risk management (JCHS, 1993). These principles were also recommended by a recent report of the IAEA (1995).

> Principle 1 Decisions for the public in regard to health and safety must be open and apply across the complete range of hazards to life and health.
>
> Principle 2 Risks shall be managed to maximize the total expected net benefit to society.
>
> Principle 3 The safety benefit to be promoted is life-expectancy.

These principles are consistent with the proposed criteria. The first principle is very important - that the marginal cost of saving a life in environmental protection should be the same as the marginal cost of saving a life in manufacturing, or the cost of saving a life in health care. Moreover this should be clearly visible to the public and other stakeholders because of the "open" nature of decision making.

The Canadian committee (JCHS, 1993) proposed a QUALY measure (Quality Adjusted Life Years) rather than simple life expectancy selected by IAEA, as the measure of safety (or risk reduction). However, because of the high degree of correlation between the quality of life elements in HDI and life expectancy, there is little practical difference between the two criteria. There are a number of case studies presented in the IAEA document which demonstrate the practical use of this criteria in risk decisions.

The second principle states that it is the net effect to society that is important. The costs of the risk management controls must be deducted from the savings of lives and the indirect costs and benefits to society at large must be accounted for. This requires consideration of safety benefits of economic development in the risk decision. The importance of this consideration was demonstrated in a study of the importation of hazardous wastes in West Africa (Asante-Duah, 1992) where the expected result was "no importation of wastes" (as supported by the Basel convention). This was not the result when the life-expectancy impacts of the economic activity were included. When indirect economic impacts on life-expectancy are included the importation of hazardous wastes is potentially a desirable activity for West Africa, conditional on the reliability of management controls for the direct risks involved in the transport and processing of the wastes.

IMPLICATIONS FOR INNOVATIVE TECHNOLOGIES

If it is accepted that the most humane approach to environmental protection is a performance based HDI criteria, this leads to different levels of protection in different countries. There would need to be a variety of techniques developed for used in different countries. In each case the objective would be to develop a set of techniques that are the most efficient at a number of different levels of environmental protection suitable to the country in question.

For example, in doing clean up of contaminated soils, the Thermal Desorption technique is often much less expensive than Incineration but generally can not meet existing clean up standards. With performance based criteria it may be possible to use this technique to achieve 95% of the protection level at a fraction of the cost. The use of bioremediation is another example of technologies that could be utilized in a performance based approach.

In general, with a performance based approach there would be an increased demand for innovative technologies that were cost effective (in a life expectancy sense) at lower levels of environmental protection. Environmental technologies used in different countries would be different.

FRAMEWORK FOR RISK DECISIONS

Figure 3 gives a framework for risk decisions. The proposed HDI criteria is represented as the "public interest". In most cases, the public interest is not directly connected to risk decisions. If it was considered desirable then the proposed HDI performance criteria could be connected through regulations.

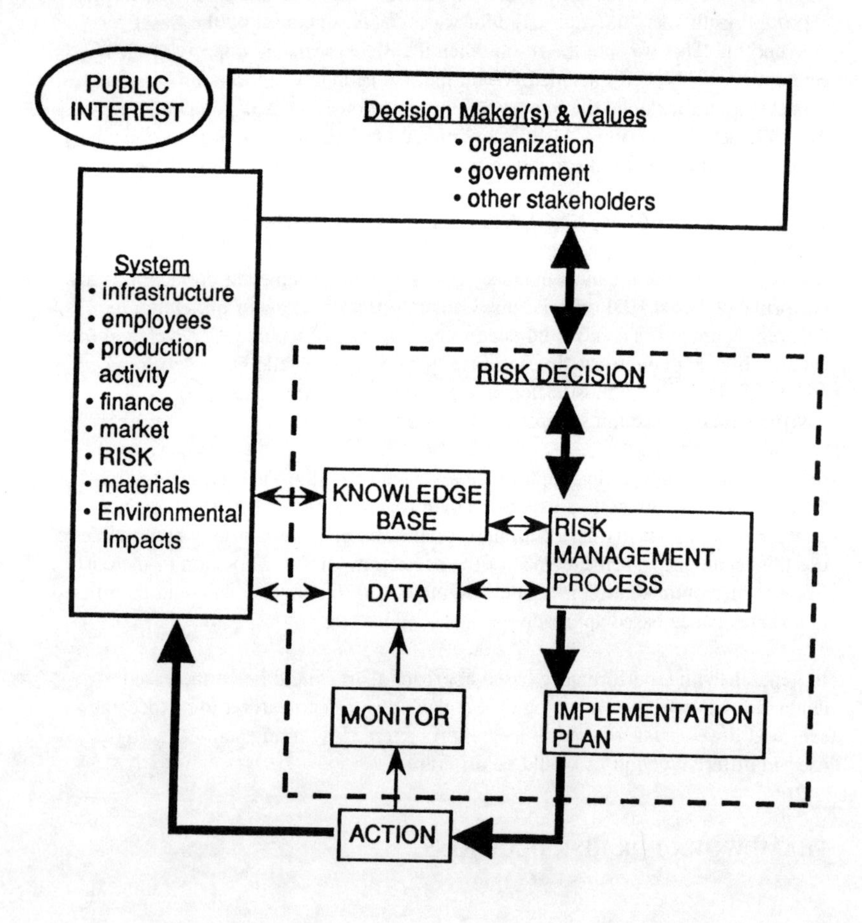

Figure 3 Framework for Risk Decisions

The "system" in Figure 3 for environmental protection is the whole assembly of the natural environment, buildings, administration, material inputs, production activities, product outputs, etc. of a firm or organization, including the associated risks of air, soil, and water pollution. The system inputs into the risk decision/risk management process are represented by "knowledge base" and "data".

For environmental protection, the usual situation is a private firm or institution making decisions under regulatory control. The government regulator is shown as a "decision maker" or critical stakeholder. In this role the regulator has considerable decision making powers, especially under performance based criteria. The purpose of Figure 3 is to outline the key elements in the decision framework, one of these elements is the Risk Management Process.

Risk management is a process to assist decision maker(s) and it has direct and continuous interaction with the decision maker(s) (CSA, 1996). The "actions" taken (e.g. clean up of stack emissions, reduction in waste materials, introduction of new manufacturing processes using biodegradable materials, and other environmental protection or risk controls) are based on the risk management process and the decisions taken are "monitored" by collecting data before and after the action to be sure the risk controls break even.

Figure 3 outlines a decision making environment that is clearly a cyclical process and is in fact a continuous improvement process. This is not unexpected since environmental protection is one of the dimension of quality. Environmental risk management is done by line managers in conjunction with cost reductions, improvement in customer satisfaction, development of new products, and all other quality aspects of Total Quality Management.

The Canadian risk management process will be discussed next since it is a key process at the next level down from the risk decision framework shown in Figure 3. At a third level down there are tools and techniques of risk analysis, hazard identification, risk communication, measurement of stakeholder preferences, etc.

At the level of decision making and throughout its elements the role of judgment is paramount. Data is usually incomplete and does not identify unambiguously cause and effect relationships in environmental damage or protection. Much of the useful data is contained in the knowledge base resident in people who have acquired it through experience with the system, training, and consultation with others. Not only decisions but analysis, and interpretation of results depends heavily on expert judgments.

Knowledge base in risk management is represented by the people on the risk management team plus the stakeholders who are consulted, including government regulators. They in turn may access knowledge in universities, the literature, data in research laboratories, results of human factor experiments, and so forth. The understanding of the environmental risk scenarios, potential risk control options and the evaluation of the "best" control option all depend on the knowledge base. In most cases, the knowledge base is contained in the "experience" people have with the system and its operation.

One of the objectives of training is to enhance the available quality of the knowledge base in the organization. This must be done at all levels in the organization: operations, design and planning, and direction of the organization. Too often, the development of a knowledge base at higher levels in an organization is forgotten. Special knowledge is required for risk decisions as well as for operations.

For example, in Canada there are relatively few people with knowledge in the risk management process. This has resulted in decisions on risk that ignore stakeholders' concerns. Ontario recently spent $CAN 160 million on the siting of a waste treatment facility that was never constructed. The possibility of "no acceptable location" was not included in the data for the risk decision process. Yet there is considerable literature and experience with siting decisions that identifies this issue and provides mechanisms for resolving it, without the expenditure of large amounts of money.

The knowledge base available will limit the rate at which improvements can be made in the risk decision activity. The rule of thumb in Canada is that changes in safety culture will take from 3 to 8 years to achieve. Most of the inertia in the system is related to the required change in the knowledge base. If people are convinced that they have to change their approach in order to reduce risks; then, it takes time for them to learn how to do environmental risk management; it takes time to gather data and analyze it; it takes time to develop and evaluate risk control options; and it takes time to take action. Moreover, in most firms this process must take place sequentially at different levels in the hierarchy: first the directors, then the managers, then the supervisors, then the support staff, and then the operators.

It is also instructive to consider the time required to build a knowledge base for chemical safety in the UK. From an initiating event in 1974 (Flixborough) the development of a knowledge base is clearly marked by milestone reports every 2 or 3 years, continuing even today. The result so far is a comprehensive and effective system. It is not clear how this process could be accelerated and in fact I am not aware of anyone even studying the successful development of this knowledge base with a view to making improvements in other areas that are not so advanced.

The knowledge base is the source of expert advice and judgments in ALARP (As Low As Reasonably Practical) (HSE, 1989) which in turn plays a central role in risk decisions. Because of the uncertainty in risk, in effectiveness of risk controls, and most importantly in stakeholders views, the selection of the ALARP or

ALARA risk control option is normally not possible based on analysis alone. The use of expert judgment is needed. The quality of expert judgment depends on people's knowledge base.

RISK MANAGEMENT PROCESS

Down one level from the risk decision framework is one of the most important components - the process of risk management. Figure 4 gives the draft version of the Canadian Standard model for "risk management decision process" (CSA, expected 1996). Figure 4 with the text of the standard provides a step by step process that decision makers and their risk managers can follow to make, hopefully, better risk decisions. The risk management process is where factors additional to HDI (e.g., loss of natural environments, impacts of population growth, species diversity) are considered and weighed in the decision balance. The Canadian process for risk management is just one typical example of a structured stakeholder approach to decision making.

The use of a standard and structured approach is essential in risk decisions because of the uncertainty of risks and risk controls. Risk is inherently uncertainty (i.e., the probability of a loss). Risk is also uncertain because of perceptions of risk, interaction of stakeholders, etc. As a result it is almost impossible to be certain about the right thing to do. Risk decisions do not stand up well in court if the question is "Was the decision correct?" but they do stand up if the question is "Was the decision made correctly?".

The Canadian process in Figure 4 includes an Initiation step and an Action step which have already been discussed under the risk decision framework, one level up in the hierarchy (i.e. Figure 3). They are included here as links to that higher level. The Canadian standard has a number of innovations, one of these is the central theme that risk management's function is to assist and support decision makers. This is reflected in the addition of the Initiation and Action steps, which are normally not shown in models of the risk management process.

The Preliminary Analysis and Risk Estimation steps incorporate the traditional "Hazard Identification" and "Risk Analysis" activities. Figure 5 illustrates the elements of these activities as set out in the 1991 Canadian Standard (CSA, 1991). Figure 5 represents the next level down in the hierarchy of environmental risk decisions - the level of tools and techniques for implementing the process of risk management. Figure 5 will not be discussed in this paper, but as indicated the standard provides detailed clauses and advice for carrying out environmental risk analysis. For example, the standard requires the use of ranges of risk estimates to reflect explicitly the uncertainty of risk and specifies requirements for documentation and verification.

RISK MANAGEMENT DECISION PROCESS

Risk Communications

Initiation
- Define the problem or opportunity and associated risk issue(s).
- Identify risk management team.
- Assign responsibility, authority, resources.
- Identify potential stakeholders.

Preliminary Analysis
- Define scope of the decision(s).
- Begin Stakeholder Analysis (S.A.).
- Begin to develop Risk Information Base.
- Hazard identification using risk scenarios.

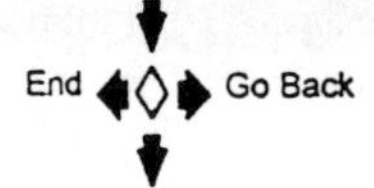

Next Step and/or Take Action

Risk Estimation
- Estimate frequency of risk scenarios.
- Estimate consequences of risk scenarios.
- Refine Stakeholder Analysis through dialogue.
- Update Risk Information Base.

Next Step and/or Take Action

Risk Evaluation
- Team meets to integrate Risk Information Base data.
- Estimate costs and benefits.
- Assess stakeholder acceptability of risk.

End ◊ Go Back

Next Step and/or Take Action

Risk Control & Financing
- Identify feasible risk control options.
- Evaluate control options in terms of risks, benefits, and costs.
- Assess stakeholder acceptance of residual risk.
- Evaluate options for dealing with residual risk.
- Assess stakeholder acceptance of proposed action(s).

End ◊ Go Back

Take Action

Action
- Implement chosen control, financing & communication strategies.
- Team evaluates effectiveness of risk management decision process.
- Establish ongoing monitoring process.

Figure 4 The Proposed Canadian Process for Risk Management (This is still a Draft of the Standard)

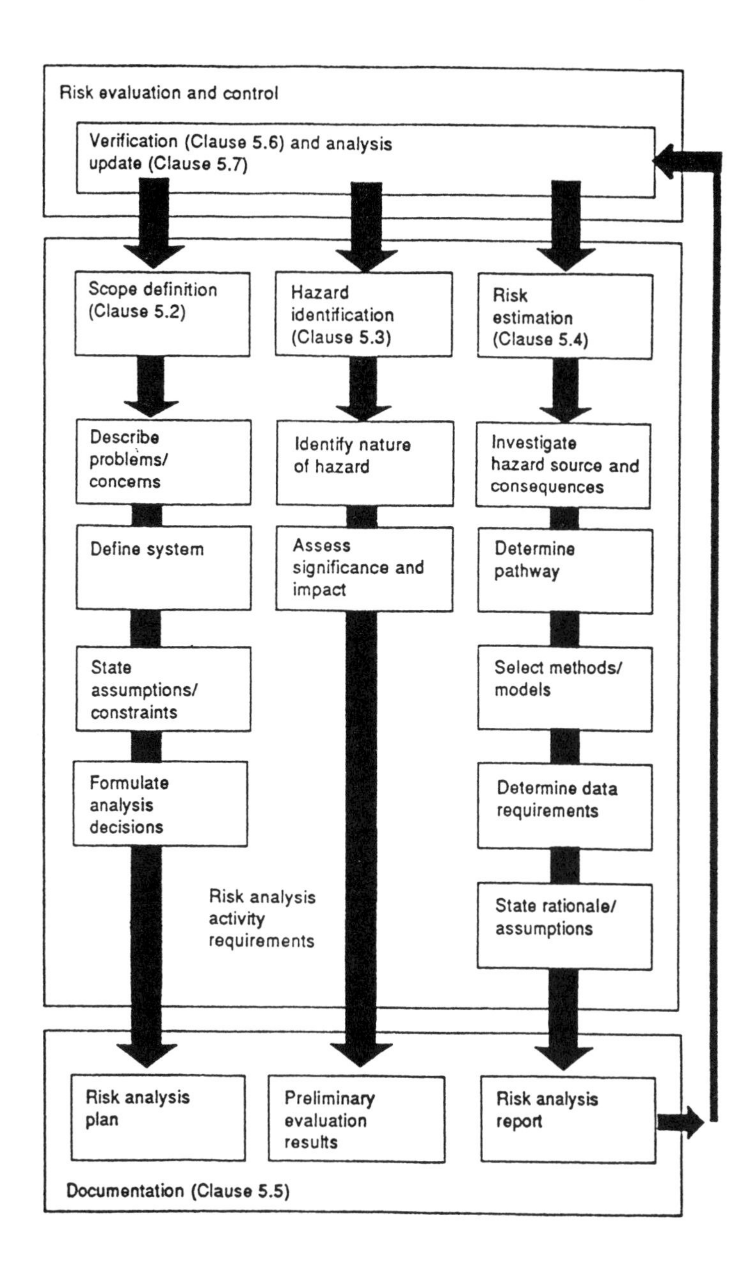

Figure 5 The Canadian Risk Analysis Process

In Figure 4, the 1996 Canadian Risk Management Standard has explicitly added stakeholders and defined them as "anyone who is affected by or believes that they are affected by a decision". The standard requires an iterative approach to identifying stakeholders and their "Needs Issues and Concerns" (NICs). In the Preliminary Analysis step the stakeholders NICs are hypothesized, but at the Risk Evaluation step they would be measured in a defined consultation program, provided that the decision maker decides that that is necessary.

The Risk Management Standard explicitly introduces requirements for Risk Communication both with stakeholders and with people inside the organization. For environmental risks, most of the stakeholders would be external to the organization and extensive risk communication would be required at all steps in the risk management process.

The Decision between steps in Figure 4 consists of four possible decisions which are made on each of the identified risk scenarios. The possible decisions are:

1. End: The risk scenario be set aside because the risk is too small or there is no known way to manage the risk scenario;

2. Action: There is a standard control for the risk, or it is unlikely that further analysis will change the action to be taken, or action is required immediately to satisfy stakeholders;

3. Back: Go back a step to update information on risks, stakeholders, or decision makers;

4. Next: Continue to the next step in the process.

The decision is made by the decision maker or may be delegated to the risk management team. The standard requires the decisions to be explicit and to be documented. This decision approach is similar to that followed by HAZOP, FMEA, and many other techniques for Hazard Identification. Since decisions are related to risk scenarios there can be a mixture of the four decisions at any step.

Risk Scenarios are the "line accounts" of risk management. Analogous to accounting, where a cost must be assigned to a line item in the budget, in risk management any failure, accident or loss must be assigned to a risk scenario. For example, "pipe fails with spill of oil into waterway and subsequent contamination of water supplies" would be a risk scenario that would be identified with all cases of spills of oil into a waterway. The management team assembled to carry out the risk decision activity develop the risk scenarios in the hazard identification task.

Risk scenarios are usually "plotted" on a Consequence-Frequency diagram according to the order of magnitude of the risk. Figure 6 illustrates a typical

Consequence-Frequency diagram used in the Canadian Oil Production Industry (Atkinson, 1995). The descriptions of the frequency and consequences are varied according to the type of loss as illustrated. The shading in Figure 6 represents the pre-assigned risk levels as a cell in the Consequence- Frequency Diagram. In other methods such as the Zurich Hazard Analysis method of hazard identification the management team assign risk levels as a part of their activity since they vary with the scale of the risk decision.

In Figure 6 the dark areas are risks scenarios that are eliminated, the grey areas are risk scenarios subjected to control (ALARP), and the white areas are risk scenarios that usually are not controlled further.

Risk Evaluation is the step where the risk scenarios are evaluated in terms of the acceptability of the risk (if acceptability or tolerability standards are available), the benefits from the activity associated with the risk (including the indirect life-expectancy benefits due to economic development), the likely costs involved with controlling the risk, and the NICs of the stakeholders. After the risk evaluation step a decision is made to control or not control the risk. The risk is either "acceptable" and the decision is "end" or the risk proceeds to the control step where ALARP or similar principles will be used to evaluate the control that is As Low As Reasonably Practical. It is at this level that the HDI criteria would be evaluated in terms of other considerations.

Risk Control in Figure 4 is the step that develops risk control options, estimates the risk reduction due to the control and evaluates the control option. This step takes the "risk with control" back through risk estimation and risk evaluation. The evaluation also includes the costs, benefits and risks associated with the controls themselves.

The risk control activity terminates in a decision to take action on a control, go back to a previous step to consider other control options (e.g., more effective in reducing risk or meeting the NICs of stakeholders), or to do nothing since it appears that the risk can't be controlled effectively.

The process described in Figure 4 has been used extensively by the author for diverse issues: transport of hazardous wastes; environmental impact of battery disposal; transportation of oil in the Arctic; occupational health; safety of the blood supply system; consumer product safety; contamination of the food chain in the Arctic; etc. It has proven to be a robust and useful process for guiding the risk management activity. It should be noted that the process is very similar to many others used around the world.

1 POTENTIAL CONSEQUENCES (For any incident check all effects)

CONSEQUENCE CATEGORY	CONSIDERATIONS			
	HEALTH/ SAFETY	PUBLIC DISRUPTION	FINANCIAL IMPACT	ENVIRONMENTAL IMPACT
I	Fatalities/serious impact on public	Large community	Corporate	Major/extended duration/full-scale response
II	Serious injury to personnel/limited impact on public	Small community	Division	Serious/significant resource commitment
III	Medical treatment for personnel/no impact on public	Minor (families)	Department	Moderate/ limited response of short duration
IV	Minor impact on personnel (first aid)	Minimal to none	Other	Minor/little or no response needed

2 PROBABILITY

PROBABILITY CATEGORY	DEFINITION
A	Possibility of repeated incidents
B	Possibility of isolated incidents
C	Possibility of occurring sometime
D	Not likely to occur
E	Practically impossible

3 RISK ASSESSMENT MATRIX

CONSEQUENCE	PROBABILITY				
	A	B	C	D	E
I	1	2	5	6	10
II	3	4	7	11	15
III	8	9	12	16	18
IV	13	14	17	19	20

Figure 6 Example of Frequency Consequence Diagram and Resulting Risk Classes

CONCLUSIONS

1. The choice of environmental protection levels and the selection of environmental technologies should be based on HDI (UN Human Development Index) as a measure of sustainability.

2. It is expected and desirable that there would be different levels of environmental protection in different countries in order to reflect a common concern for attaining the maximum quality of life.

3. Existing methods for setting environmental protection levels in Canada are currently based on a weight of evidence, performance criteria. Moreover, the decision process used is similar to that proposed in the Canadian Risk Management Standard.

4. For effective environmental risk management it is essential to have a process that accommodates all stakeholders and has extensive risk communication. The role of value judgments must be explicit.

5. Analysis of environmental risks should explicitly reflect uncertainty.

6. The use of risk management techniques in conjunction with a HDI performance based criteria is practical and feasible.

REFERENCES

Atkinson, D., 1995 "Hazard Identification and Risk Analysis for Oil and Gas Production Facilities", PSLM 1995 Conference Pre-print - Toronto, Institute for Risk Research, Waterloo, Canada.

Asante-Duah, D., Saccomanno, F., and Shortreed, J., 1992,"The Hazardous Waste Trade: Can it be Controlled?", Environmental Science and Technology, Vol. 26, No. 9, American Chemical Society.

CSA, (Canadian Standards Association), 1996(proposed), "Risk Management: Guidelines for Decision-Makers", CAN/CSA-Q850, Quality Management a National Standard of Canada.

CSA, (Canadian Standards Association), 1991, "Risk Analysis Requirements and Guidelines", CAN/CSA-Q634-M91, Quality Management a National Standard of Canada, Toronto.

HSE, (Health and Safety Executive), 1992, Management of Health and Safety at Work: Approved Code of Practise, HMSO, London, U.K.

HSE, (Health and Safety Executive), 1989, Quantified risk assessment: Its input to decision making , HMSO, London, U.K.

IAEA, (International Atomic Energy Agency), 1995, "Principles and Recommendations for the Integrated Management of Technological Risks", Working Material, Vienna.

JCHS (Joint Committee on Health and Safety of the Royal Society of Canada and the Canadian Academy of Engineering), 1993, "Health and Safety Policies: Guiding Principles for Risk Management", Institute for Risk Research, Waterloo, Canada.

Lind, N. C., J.S. Nathwani, and E. Siddall, 1992, Managing Risk in the Public Interest, Institute for Risk Research, Waterloo, Canada.

Paustenbach, D.J., 1995, Retrospective on U.S. Health Risk Assessment: How Others Can Benefit, Risk, Health Safety and Environment, Vol. 6, No. 4, pp283-332, Franklin Pierce Law Center, Concord, NH.

Siddall, E., 1989, A study of Mortality in Canadian Cities, Paper No. 13, Institute for Risk Research, Waterloo, Canada.

Shortreed, J.H., R. Bennett, and W. Cherry, 1992, Benchmark Risks, in McColl, S. (ed), Development of Environmental Health Status Indicators, Institute for Risk Research.

UNDP (United Nations Development Program), 1995, "Human Development Report 1995", Oxford University Press.

World Health Organization, 1992, "Our Planet Our Health: Report of the WHO Commission on Health and Environment", Geneva .

Integration of Biotechnology in Remediation and Pollution Prevention Activities

Janet M. Strong-Gunderson*

> ***"Convinced of the importance of the conservation, protection and enhancement of the environment in their territories and the essential role of cooperation in these areas in achieving sustainable development for the well-being of present and future generations ..."***
>
> *preamble to the North American Agreement on Environmental Cooperation*

The *North American Free Trade Agreement/North American Agreement on Environmental Cooperation* provides a mechanism for an international collaboration between the United States, Canada, and Mexico to jointly develop, modify, or refine technologies that remediate or protect our environment. Our countries have a vested interest in this type of collaboration because contaminants do not respect the boundaries of a manufacturing site, region, city, state, or country.

The Environmental Sciences Division (ESD) at Oak Ridge National Laboratory (ORNL) consists of a diverse group of individuals who address a variety of environmental issues. ESD is involved in basic and applied research on the fate, transport, and remediation of contaminants; environmental assessment; environmental engineering; and demonstrations of advanced remediation technologies. The remediation and protection of the environment includes water, air, and soils for organic, inorganic, and radioactive contaminants. In addition to remediating contaminated sites, research also focuses on life-cycle analyses of industrial processes and the production of green technologies. I will focus this discussion on subsurface remediation and pollution prevention; however, our research activities encompass water, soil and air and many of the technologies are applicable to all environments. The following discussion focuses on the integration of biotechnology with remediation activities and subsequently linking these biological processes to other remediation technologies.

* Building 1505, Mail Stop 6038, Environmental Sciences Division, Oak Ridge National Laboratory, Oak Ridge, TN 37831-6038 USA

Environmental Remediation/Risk Assessment: The large number of contaminated sites throughout the North American continent necessitates the evaluation and ranking of these sites according to contaminant concentration and type, extent of both vertical and horizontal contaminant dispersion, and the potential immediate and long-term impact to human health and the environment. The evaluation must take into account the current state of technology development and the finite funding available for remediation activities. This need for ranking sites does not suggest that all sites must be evaluated prior to initiating cleanup activities. It will be relatively simple to select a site or sites that have the highest immediate human and environmental impact/threat and can be remediated through the use of current technologies. Risk analyses groups within ESD are involved in these types of activities.

The sites chosen for the initial collaborative remediation will require site characterization and risk assessment to provide the information needed to design a remediation plan and time frame. Community support is very important; therefore, choosing sites that can be cleaned up with current technologies not only increases public acceptance and confidence but also provides full-scale evaluation of many current pilot-scale technologies. Furthermore, these relatively simple demonstrations will provide invaluable field experience to further development of new technologies that are required for the more difficult and unique sites.

Most contaminated sites can benefit from a combination of physical, chemical, and biological, remediation methods. Research within the ESD Microbial Interactions Group has focused on bioremediation activities combined with physical treatments, such as deep soil mixing and vapor extraction and chemical treatment through the use surfactants.
The specific topics we are examining include:

1) the selection of bacteria capable of degrading contaminants;
2) enhancing the bioavailability of contaminants for microbial degradation through the use of surfactants and biosurfactants;
3) monitoring the distribution of bacteria, nutrients, surfactants, etc. by using classical microbiological methods and innovative, real-time tracers; and
4) using engineered microorganisms as bioreporters to measure the physiological status of the microbial community during remediation activities.

The simultaneous investigation into these parameters as well as a close collaborations with engineers, geologists, chemists, and hydrogeologists minimizes the risk of failure.

The remediation strategy will be designed on the basis of site assessment, environmental impact analyses, and previous experience. Recommending natural contaminant attenuation (intrinsic bioremediation) may be possible and economically feasible. This treatment involves monitoring the contaminant concentration levels in combination with the natural bacterial populations for their ability to degrade the contaminants with limited to no input, such as nutrients and oxygen. If natural attenuation is not possible or does not proceed at an acceptable rate, the addition of bacteria (i.e. bioaugmentation) could significantly increase the rate of contaminant degradation.

Bacteria capable of degrading a variety of contaminant compounds have been isolated from both pristine and historically contaminated sites (water, soil, and air), and their capacity to degrade a variety of contaminants (organic and inorganic) has been measured. Research has shown that under defined conditions nonindigenous bacteria can outperform indigenous microorganisms. The bacteria used in amendment technologies are currently not genetically engineered. However, the eventual approved use of genetically engineered bacteria in in-situ bioremediation has the potential to significantly increase the rate of contaminant degradation and thus, decrease the time required to bring a site to closure.

In addition to bacterial and nutrient/oxygen amendment, the rate of contaminant degradation can be further enhanced by the addition of surfactants. These surfactant compounds can be used to (1) increase the rates of in situ bioremediation and (2) increase contaminant solubilization for soil washing followed by aboveground treatment (e.g., bioreactors). Our recent laboratory work has shown that adding low concentrations of surfactant can enhance the overall rates of in situ bioremediation while keeping the costs at a minimum. High surfactant concentrations, although useful as a soil washing technology, can reduce the contaminants ability to be degraded by microorganisms as well as being more costly.

In addition to examining contaminant degradation by a variety of microorganisms, we have also used microbial bioreporters to directly measure the physiological processes of contaminant degradation and have used molecular methods to monitor the bacterial population active in contaminant degradation. These bioreporters provide a real-time analysis of the overall contaminant degradation rates. Thus, if oxygen becomes a limiting factor or the contaminants are not available for degradation, the system can be immediately adjusted to maintain optimal operating (i.e., degradation) conditions.

These bioreporters have also been used as an efficient and inexpensive screening tool to detect contaminants within groundwater samples. The bioreporters can provide a real-time analysis of a waste stream to determine sporadic leaching or accumulation of volatile organic compounds. Through these research activities, we are integrating basic microbial ecology with molecular biology, which enables us to design and optimize remediation regimes that are both cost effective and efficient.

Pollution Prevention/Life Cycle Analyses: In addition to in situ contaminant remediation, research has also focused on the use of microorganisms and bioreactors in pollution prevention and waste minimization technologies. Technologies for recycling chemical compounds at the industrial sites can not only reduce operating costs but can be used to enhance public relations. If recycling is not a viable or economically feasible option, degradation of the waste stream on-site can reduce disposal costs and prevent the generation of potential, future pollutants.

Information gained from in situ bioremediation experiments/demonstrations (discussed above) has enabled the design of unique bioreactors that operate as end-of-pipe technologies. Bacteria isolated from previous bioremediation and site characterization studies have been used as inoculum in bioreactors and in land-farming activities. These activities occur on-site and have the potential to significantly reduce environmental impacts because they do not generate hazardous waste. Furthermore, they increase positive public relations between the communities and industry.

Collaborations/Site Closure Levels: In addition to other work within the United States, our group has recently expanded research activities to include unique contamination problems in the south of Mexico and in the northern regions of the United States and Canada. The work in Mexico is focusing on the remediation of weathered crude in regions with high environmental temperatures. These temperatures are not typical in the United States; thus, this project is a collaboration between the United States and Mexico to jointly develop and modify existing technologies for use at these sites. In addition to high temperatures, low temperature are also an area of interest. We currently have several projects in the United States and Canada that are examining the degradation of glycols in the subsurface as pollution prevention activities.

Regardless of the contaminated matrix (air, water, subsurface) or location (United States, Canada, Mexico) the ultimate goal is to reach a cleanup level that permits site closure. In many cases, this level is not standardized between states let alone between countries. It will be a particular challenge to establish acceptable remediation levels and measure them accurately; keeping in mind the costs and balancing that with the level of immediate human and environmental impact.

Oak Ridge National Laboratory is Managed by Lockheed Marietta Energy Systems, Inc., under contract DE-AC05-84OR21400 with the US. Department of Energy. This research was sponsored by the In-Situ Remediation Technology Development Program of the Office of Technology Development (J. Walker, Program Manager). Environmental Sciences Division Publication.

Water Supply and Wastewater Treatment/Reuse

Timothy J. Downs*

1. Opening Discussion

The President of Mexico has promised to address the problem of wastewater in Mexico City. This makes the issue a priority. The current treatment plan is being promoted by the National Water Commission (Comisión Nacional del Agua - CNA) and involves what is termed "advanced primary" treatment using sedimentation with the application of aluminum sulfate to kill *helminth eggs* which represent a demonstrated health risk. The recommended design is based solely on this microbiological contaminant criterion because such data are available and helminths have a demonstrated health impact on the rural population that use the wastewater for irrigation. The plan is about to be implemented, despite controversy about the management of the large amounts of sludge that will be generated (the average flow of wastewater is 40 m^3/s) and unanswered questions about chemical contaminants.

This context provides background for a more general discussion. Three water uses exist: agricultural, industrial and domestic. A loop such as **Figure 1** shows one water use-wastewater loop. For the three uses we require three flow quantities, three flow qualities, and three treatment processes and costs. For sustainable water and wastewater management we must begin to apply these loops.

The focus tends toward urban areas. However, in rural Mexico where more than half the country's population live in conditions of poverty, acute water shortages, lack of safe water and sanitation, and river pollution from pesticides and animal wastes are serious problems.

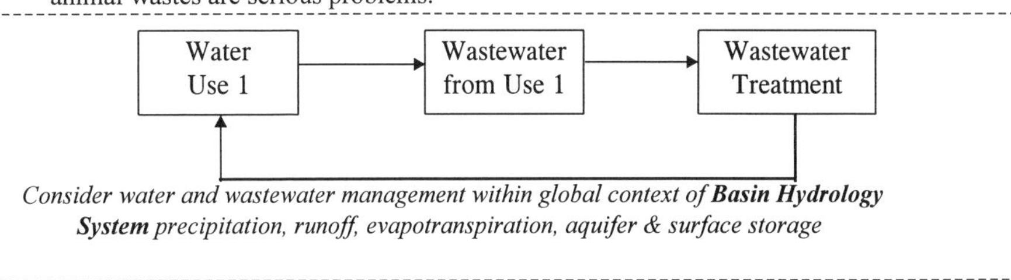

Figure 1. Idealized Closed Loop Sustainable Water Supply & Wastewater Treatment

* Instituto Nacional de Salud Pública, Ave. Universidad 655, Col. Sta. Ma. Ahuacatitlán, Cuernavaca, Morelos, CP 62508 México

2. Priority Problems and Challenges - Brainstorming

The following topics were mentioned as problem areas (unranked):
1) low cost treatment for animal waste treatment; 2) rainfall aquifer re-charge technology; 3) pesticide residues sorption to soils; 4) cost efficient technology for Mexico (Best Appropriate Available Technology - BAAT); 5) research to understand complete water situation with hydrological framework (Figure 1); 6) infrastructure cost for reclaimed water; 7) policy + economic incentives for reuse; 8) priority information for decision makers for reuse planning; 9) recharge area protection and non-point source pollution; 10) human resource formation - training; 11) training of professionals in countries with experience e.g. Israel, Germany (Rhine River), Switzerland, South Africa; 12) toxic chemicals fate and transport; 13) waste minimization; 14) feasibility of reuse in low income residential areas and high income areas - education about reuse technology, especially long-term costs of no reuse; 15) education of professionals in general; 16) water supply and reuse cycle; 17) enforcement of regulations; 18) increasing awareness in Tourist Industry (and Industry in general) of simple, economic conservation measures; 19) impact of tourism on water quality; 20) source inventory and source control technology; 21) research needs for technology; 22) cleaning groundwater using micro-organisms isolated from the deep subsurface; 23) reuse? under what conditions can it be incorporated into water supply? - a big problem also in the USA; 24) water losses and efficient use of water; 25) water metering; 26) source control and migration control for the following contaminants: heavy metals, chlorinated solvents, organics, phenols, bacteria, virus and other pathogens e.g. helminths, hydrocarbons, detergents (non-biodegradable), PAHs, PCBs, pesticides and fertilizers. 27) Standards? What are the US standards? The standards vary by State and EPA has federal standards. Example values and criteria are: organic matter, nitrogen (ammonia), pathogens (*e-coli*) and sometimes phosphates, BOD 5-30 mg/l, COD TOC 10-50 mg/l, nitrogen: ammonia 1 mg/l, nitrate 5-10 mg/l, phosphorus <1 mg/l, pathogens 1/100 ml, residual chlorine 0.1-0.5 mg/l. In Mexico standards are being re-organized into 3 main ones and discharge conditions will be applied depending on the "assimilation capacity" of the receiving water body (research into *assimilative capacity* is needed - frontier ecological systems research: what criteria? what indicators?). Current Mexico standards are: - BOD 30-120 mg/l range, N or P not used, pH, oils and grease, coliforms, but the norms do not follow a rational line according to some professionals.

The Priority Problems and Challenges were finally organized into five subgroups: Technologies, Reuse, Training, Regulations, Characterization.

3. Research Lines - Further Brainstorming

The following research lines were identified:

1) We must consider wastewater as part of the hydrologic cycle at its point of reuse: downstream reuse, upstream reuse, secondary use, aquifer recharge, or simply water body discharge. Example: water supply in Atlanta (3 million people) where the river goes to Alabama, Florida and pollution impacts bays, and issues

of river and aquifer recharge are 'hot'. People talk of 'water wars' and Atlanta is not slowing down in growth;

2) Mexico City has similar use issues and we do not have good knowledge of all contaminant levels;
3) in Latin America public and political conscience about water is very different to the USA and CAN - it is accepted that in Mexico water is not fit to drink;
4) taking a watershed, we may ask: how can you optimize the system?
5) the National Water Plan of Mexico focuses on Border Cities and Mexico City, while ignoring a vast rural problem area;
6) context for development of a water plan is needed e.g. USA criteria are that the water body be fishable or swimmable - the state of Mexico's rivers is that many are used as sewers and while the Mexican standards are the "best" in Latin America, they are not enforced;
7) political consciousness is key - it is why Israel water policy works;
8) understand national differences between and within Mexico, USA and Canada;
9) community driven plans, especially for rural areas are needed, for example plans based on indigenous culture;
10) research into primary treatment technologies: - common in Mexico is activated sludge, in the USA and CAN primary and enhanced primary + biological is used;
11) oxidation ponds, stabilization ponds at appropriate scales;
12) sludge stabilization is a relatively new problem in the USA for land disposal and application, but food chain impacts remain to be investigated.
13) innovative technologies for marginal areas, rural areas, portable systems;
14) cost effectiveness of solar energy for treatment?
15) bio-removal of heavy metals (phyto);
16) warm climates anaerobic treatment limits;
17) phyto-remediation (breakdown of organics to carbon dioxide, water and inorganic components), e.g. using constructed wetlands;
18) phyto-accumulation of metals and radionuclides e.g. secondary recovery of useful mineral waste (lead, gold, uranium) using plants.

4. Summary

The following was presented as a summary of requirements for further work:

◊ *Technologies:* Low-cost treatment for animal wastes; aquifer recharge; best available appropriate technologies; infrastructure development; training in biotechnology; bio-systems - algal, phyto-remediation, microbial systems; low-income area waste treatment; source inventory and source control (water and contaminants).

◊ *Characterization:* Flow sources - hydrology; waste streams: heavy metals, chloro-solvents, BOD, phenols, virus, pathogens e.g. helminth eggs, hydrocarbons, non-biodegradable detergents, pesticides, PAHs, PCBs, radionuclides, nitrogen (NH_4, NO_3), phosphorus. Characterization of receiving water bodies and ecological systems - “assimilative capacity”; environmental impact of current practices, full contamination accounting.

◊ *Regulations, Laws and Oversights:* Enforcement of laws; water use controls; assessment of and use of natural attenuation; pollution prevention; waste minimization; source control; conservation; connection of water supply and reuse water.

◊ *National Water Quality Plans:* Major differences between CAN, USA, Mexico; critical review and development of appropriate plans including economic issues and community involvement.

◊ *Conventional Treatment Technologies:* Activated sludge (organics); removal; pre-removal; primary and enhanced primary treatment; sludge disposal and stabilization (aerobic, anaerobic); disposal at sea and on land.

◊ *Innovative Technologies:* Scaled oxidation and stabilization ponds; small self-contained systems; solar-driven O_3/H_2O_2 systems; high rate oxidation ponds; UV disinfection; collection systems; bioaccumulation of heavy metals - phyto-accumulation of heavy metals including secondary recovery of valuable metals (gold, uranium, zinc, lead); limitations of anaerobic systems for sewage treatment; phyto-remediation of organics e.g. constructed wetlands; bioremediation.

◊ *Reuse:* Policy and economic incentives for reuse; information to prioritize reuse options; recharge and protection options; reuse connection of water supply and wastewater treatment; natural attenuation of contaminants; efficiency of use.

5. Interfaces

Many common interests exist between the Water Supply and Wastewater Treatment/Reuse Area the Bioremediation Area and the Groundwater Area. The common interests are:

- Site characterization and monitoring
- Fate, biotransformation and transport
- Natural biochemical attenuation processes
- Treatment technologies and applications
- Training courses

In conclusion, it was recognized that only by working in a country for a reasonable period of time is it possible to appreciate that country’s specific problems and needs. Professional exchanges between Mexico, the USA and Canada must be encouraged to foster technical and cultural understanding.

Geo-Environmental Concerns in North America

Emir José Macari*

This paper is a summary of the breakout discussion sessions that took place during the "Geo-Environmental Issues Facing the Americas," Workshop in Mayagüez, Puerto Rico in September 1993. The proceedings from this workshop were published by the American Society of Civil Engineers, Geotechnical Special Publication 47, ASCE, 1994.

Environmental Regulations

One should first focus upon the differences and similarities between Geo-Environmental regulations in developing and developed countries and the regulatory environment. Hence one should stress the differences between performance based regulations of the United States and how these have evolved from the original format of prescriptive based regulations of the 1970s. However, Latin America prescriptive regulations are currently the norm and few efforts are being directed at changing this philosophy. These regulations are affected by a number of factors which include local perception of risk (cost/benefit) as well as socioeconomic issues which vary greatly even within the developing countries. As a result, the field of environmental regulations is quite disperse in the western hemisphere. The United States and Canada have well established environmental regulations while other countries such as Brazil and Mexico have developed, over the last decade, detailed regulatory procedures in this arena. However, there are other countries in Latin America where environmental regulations are just beginning to be drafted and others, yet, where they are not even being considered.

In addition, one should also recognize that in the enforcement of environmental regulations an even wider gap exists between developing and developed nations of the Americas. The United States has substantial punitive measures to ensure that industry and government, alike, comply with present regulatory measures. However, the damages of past practices are very difficult to overcome. Currently "avoidance" is the word that must be stressed, however, large

* Associate Professor, School of Civil and Environmental Engineering, Georgia Institute of Technology, Atlanta, Georgia 30332-0355

efforts are still directed (and must be directed for many years to come) at monitoring, controlling, and remediating the damages that resulted from under-regulated economic growth of the past.

There is a strong sentiment that, in general, punitive regulatory measures will not be successful in Latin America and thus a regulatory framework which incorporated incentives could be more attractive and efficient. In addition, developing countries must learn from the mistakes of their developed neighbors and to avoid the high costs of remediation that may result from unregulated economic growth. In fact, this philosophy goes hand-in-hand with the goals of Sustainable Development which are being advocated by many nations in our hemisphere.

Environmental Site Assessment

Several issues should be stated in this section, starting with a focus on current practices in performing environmental site assessments in the USA and Canada, the differences in the need for environmental site assessments in both the developed and developing countries of the Americas, the identification and development of appropriate technologies for performing environmental site assessments in the developing countries of the Americas, impediments in using environmental site assessments as tools for protecting and restoring environmental quality in developing countries, and costs associated with the implementation of environmental site assessments in developing countries. The environmental problems of a developing country are likely to be very different than those of a more developed country. Therefore, much of the technology currently used for environmental site assessments in the USA and Canada may not be appropriate for the developing countries. In particular, the focus in the United States on using the "Best Available Technologies" is likely to be inappropriate in a developing country in many cases.

"Phase I" type screening studies which identify and rank environmental risks are likely to be more valuable than "Phase II" type sampling and testing in the developing countries of the Americas. The collection and dissemination of information on current methodologies for screening and ranking environmental risks are tasks that can be undertaken as a first step in fostering protection of the environment in the developing countries. Towards this end, one should consider the establishment of an Inter-American Training Facility, development of a data repository and on-line data base development of manuals, and technical guidance documents for environmental site assessment. While existing environmental site assessment technologies provide a good starting point for development of environmental site assessment techniques for developing countries, these technologies should be tailored to the values and priorities of the particular country in which they are to be applied.

Development of a site assessment technology database is considered to be one of the most important measures needed to facilitate implementation of environmental site assessment for environmental protection and restoration in developing countries. However, the development of low cost, reliable environmental site assessment methods for the screening of environmental problems and prioritization of remedial actions was also considered a high priority. Integrated sampling, testing, modeling, and interactive real-time methods for data acquisition, synthesis, and portrayal were identified as technologies for subsequent development to improve the effectiveness of current environmental site assessment techniques.

Ranking remedial measures by cost-effectiveness is important in developing countries, where scarcity of capital and allocation of resources are critical issues. Scarcity of resources and economic imperatives for growth and development are a major impediment to implementation of environmental site assessment in developing countries. Lack of continuity in policy and failure to communicate the importance of environmental protection and restoration are also significant impediments to the implementation of environmental site assessments.

Continuing dialogue must be established among all stakeholders - the engineering and research community, industry, the government, and the public - to establish priorities for environmental protection and remediation. Environmental hazards and risks, the relative value of environmental resources, and available remedies are likely to differ from country to country. Therefore, establishment of priorities must be tailored to the economic, environmental, political, and cultural conditions in each specific country. Failure to recognize the differences in circumstances in individual countries was discussed as a fatal flaw that would prevent transfer of environmental site assessment technology and implementation of environmental site assessment as a tool for environmental protection and restoration in the developing countries of the Americas.

Environmental Remediation

There is a growing interest in improving the quality of the environment. In the developed countries, increasing emphasis is being placed in the mitigation of environmental impacts as a vital condition for continued progress. In this sense, legislation and enforcement organizations have been established to compel the industries to reduce the contamination levels. Modern industrial technologies are focused on the optimization of their processes to minimize risks and contamination.

Several environmental problems have been found to be common in many countries in the Americas. Sites contaminated with heavy metals and/or organic contaminants, leakage from underground storage tanks, surface water contamination, uncontrolled landfills, and agrochemical contamination are some of the "shared" environmental problems. Latin American countries have only recently started to control environmental contamination. Over the last few years, in several countries, new laws and regulations have been promulgated. Consequently, remediation actions to solve environmental problems have been recently considered. However, in the Americas, due to the situation described above, the range of development and use of environmental remediation techniques are very broad. In the United States, many remediation methods are currently in use; conversely, in other countries, clean-up activities have not been widely reported.

As a first step, in the Latin American countries, contaminated sites must be identified. Second, the goals of remediation for a particular site must be defined. It is highly recommended that countries in which remediation techniques are not currently available that containment strategies for hazardous waste be developed. Containment systems would be a transitory solution for environmental problems when the sources of contamination are well located.

In these developing countries, there is a clear need for strengthening the capabilities of industries and scientific research centers to handle environmental problems. The training of personnel and dissemination of information will permit the adequate selection and use of remediation techniques, and the improvement of the regulatory framework.

Landfills and Containment Systems

Landfills and waste containment systems (LAWCS) are affected by complex engineering, socio-economic, and regulatory issues. A large disparity exists between the countries of the Americas regarding these issues. The disparity in regulations between the countries of the Americas is large. For example, the Resource Conservation and Recovery Act (RCRA) regulations governing disposal of hazardous (subtitle C) and non-hazardous (subtitle D) wastes in the USA is well established. In contrast, such regulations are nonexistent (Costa Rica) or just being established (Chile, Argentina, and Venezuela). This situation creates opportunities and provides compelling motivation for the technical communities of these countries to influence the promulgation of new regulations to best fit the local and regional conditions while using the latest technical developments, learning from mistakes that were made in the USA over the evolution of these regulations, while promoting environmental awareness and compliance.

In most instances, different regulations will be required for different categories of LAWCS. For example, regulations governing closure of existing, uncontrolled dumps will be different from regulations governing siting, design, and closure of new facilities. In addition, different countries will probably adopt different, waste specific regulations - mining, hazardous, non-hazardous, and nuclear waste - that are best suited to their conditions.

There are important relationships between technical and socio-economic aspects of LAWCS in the developing countries of the Americas. In some cases, implementation of new waste disposal regulations may severely impact the livelihood of many disadvantaged residents who live in and near landfills (e.g., Montevideo, Uruguay; Fortaleza, Brazil). Other socio-economic aspects include the institutional capacity of the regional and central government organizations to handle and enforce the new environmental regulations. The issue of private versus public ownership of waste disposal facilities is already being debated in several Latin American countries.

The criteria for new facilities in the developing countries should emphasize the importance of site selection, design standards, and construction quality assurance (CQA). These should be flexible to account for regional conditions (eg. seismicity, climate). A risk based approach to these new standards is strongly recommended. For example, seismic issues that are not a major concern for solid waste landfills, become more critical for liquid impoundments and tailings deposits.

The majority of the negative environmental impacts of waste disposal in the developing countries of the Americas are associated with many open, uncontrolled dumps. Also, new facilities will have to deal with waste screening - hazardous and non-hazardous - and waste management planning issues where the institutional capacity is generally weak.

Undoubtedly, these changing environmental conditions in the developing countries create a significant need for technology transfer (TT). The universities must initiate TT through education and research. Practitioners must obtain practical experience of the latest developments in LAWCS through their professional organizations and continuing education initiatives in cooperation with the universities. Governments must facilitate TT through education of their employees, updates and consolidation of technical guidelines and documents, and support of TT initiatives of the universities.

This workshop is an important stop in attaining cooperation among the countries of the Americas for development of improved siting, design, construction, operation, and closure standards of landfills and containment systems. The main goals of such cooperation are:

1. to identify the main country and region-specific problems associated with these facilities,
2. to increase awareness in these countries to the LAWCS problems,
3. to develop a global risk based model for siting and design of LAWCS,
4. to facilitate interaction in TT and cooperative research projects between the countries of the Americas so that the developing countries will be able to take full advantage of the tremendous experience that has been gained over the last twenty years in the developed countries.

Environmental Education

The differences among the educational and professional communities must be recognized in order to develop adequate academic strategies for cooperation among countries in North America. For example, the standard four year program in the United States is usually service-oriented, where a very broad group of the community attends and graduates from the program. Conversely, most programs in Latin America are five to six years long, and while admission requirements are relatively low, attrition rates tend to be high. As far as professional registration, it is either not required or is automatically conferred to individuals upon graduation in Latin America; whereas, the United States and Canada require professional registration, which involves professional experience and additional examination(s) after graduation.

Three essentially different alternatives - to change curricula, to place further emphasis on continuing engineering education, and to encourage team work - are recognized.

Across the Americas, changes to curricula have taken different forms. Some universities are making changes within civil engineering departments, placing special emphasis on fundamentals such as mathematics, physics, and chemistry. Some universities have organized new programs to address Geo-Environmental issues, often under the general denomination of Environmental Engineering. Curricular changes could also involve the development of special programs to adequately train technicians. There is an increased need for individuals who are skilled in field and laboratory testing, installations, and site monitoring. These changes to curricula are supported recognizing that there is need for clear understanding of the sciences in order to address the complexity of Geo-Environmental phenomena and that strengthening fundamental knowledge will enhance an individual's mobility within unstable job markets.

Continuing engineering education is the second main alternative. Short courses could take place in Puerto Rico under the CoHemis sponsorship, or they could be taught across Latin American countries, at universities, engineering societies or immediately before or after national/regional conferences. The latter alternative seems to maximize cost-effectiveness. Both students and practitioners should benefit from these courses. The committee emphasized that the design of these courses and teaching efforts should be guided by the following principles: (1) courses must be the initial step towards long-term activities rather than an ultimate objective, (2) they must stimulate technology transfer and further cooperative research, and (3) they must be oriented to problem-solving and skill-development. The following topics were identified:

- detection, monitoring, and instrumentation with demonstrations/hands on,
- design analysis and construction of containment systems
- remediation techniques such as stabilization, solidification, clean-up methods

- decision making tools such as risk management
- groundwater and contaminant migration modeling
- data basing and geographic information systems (GIS)
- Geo-Environmental aspects of geology
- presentation and evaluation of case histories
- new construction technologies and materials
- waste management planning
- special Geo-Environmental regional problems such as the effect of high precipitation, high mean daily temperature and latheritic soils.

With this in mind, the most needed short courses in Latin America are as follows: wastewater treatment, landfills, site assessment & monitoring, remediation, contaminant transport, GIS & land use, and modeling. Short courses require a modest investment (e.g., travel, local expenses, and a minimum honorarium). However, their immediate impact and the potential for initiating long-term cooperation is significant.

The solution of complex Geo-Environmental problems requires in-depth understanding of physical and chemical processes, hydrogeology, numerical modeling, geomedia behavior, modern instrumentation, local conditions and a broad environmental perspective. Such problems may be best addressed by teams of local and foreign experts who share common semantics. Unfortunately, the development of group-skills and team-aptitudes is not adequately emphasized in programs of the Americas.

Transfer of New Technologies

The term "new technology" was considered to mean technology not commercially available nor that which is currently under development. Other terms related to technology were proven, emerging, and established technology. Because the view of "new" might be different for people in various countries, "appropriate" technology was proposed as a better term (meaning appropriate for a specific location and condition). Indigenous technology was considered as technology that exists and could be adapted or transformed according to specific needs.

Appropriate technology identification and economic motivation to acquire established technologies that may be more expensive as compared to proven or emerging technologies. It is considered that, in general, technology transfer efforts should focus on appropriate technology that is cost-effective. Technology issues are related and influenced by comfort, foreign competition, and nationalism. Implementation and adaptation to different needs in different countries has to be considered to be an integral part of the transfer process. Requirements to operate a particular technology, including factors such as energy and maintenance requirements must be considered. Technical training has to be part of the technology transfer.

Current methods to learn about technology development vary. For emerging technology which is under development, information on how it is developing is typically available at conferences and in published papers. Training efforts at research labs attempt to take a technology to the next step (proven technology). When it becomes commercially available, it is considered an established technology. It should be noted that there is a potential risk when using emerging technologies under local conditions, where they haven't been proven. Alternative methods for learning about technology and technology transfer possibilities would be software exchange such as GIS, technology demonstrations on TV, and conferences. Computer networks, mailing lists, videos, and software would be media to make information and research efforts widely available and accessible. Information should ideally flow in all directions. CoHemis could focus on: information transfer having organized the data repository, using Gopher and "World Wide Web" as possible options. Information transfer at a low level, with time to develop the ideas of research partnerships and develop ideas of technology indigenous to each country, should be started. The development of expert/knowledge-based systems in environmental remediation could also aid in the transfer of technology.

SUBJECT INDEX

Page number refers to the first page of paper

AUTHOR INDEX

Page number refers to the first page of paper